Lecture Notes in Economics and Mathematical Systems

Managing Editors: M. Beckmann and W. Krelle

365

Gerald A. Heuer
Ulrike Leopold-Wildburger

Balanced Silverman Games on General Discrete Sets

Springer-Verlag
Berlin Heidelberg New York London Paris
Tokyo Hong Kong Barcelona Budapest

Managing Editors
Prof. Dr. M. Beckmann
Brown University
Providence, RI 02912, USA

Prof. Dr. W. Krelle
Institut für Gesellschafts- und Wirtschaftswissenschaften
der Universität Bonn
Adenauerallee 24–42, D-5300 Bonn, FRG

Authors
Prof. Dr. Gerald A. Heuer
Professor of Mathematics
Concordia College
Moorhead, MN 56562, USA

Prof. Dr. Ulrike Leopold-Wildburger
Professor of Operations Research
University of Graz
A-8010 Graz, AUSTRIA

Dieser Band wurde mit Unterstützung des Fonds zur Förderung der wissenschaftlichen Forschung, Wien, gedruckt.

ISBN-13: 978-3-540-54372-5 e-ISBN-13: 978-3-642-95663-8
DOI: 10.1007/978-3-642-95663-8

Typesetting: Camera ready by authors

2142/3140-543210 – Printed on acid-free paper

Acknowledgments

The work of the first author was done in part while visiting at Graz University, Austria. The hospitality and support of the University and its Institute for Statistics, Econometrics and Operations Research are gratefully acknowledged.

We express our thanks also to Professors Alex Sze and Jerry Rowell, of Concordia College, and Karl Heuer of Interactive Systems, Inc., for preparing computer programs which played a key role in discovering many of the theorems, and to Maura Cock for typing the manuscript.

Table of Contents

1. Introduction.

A Silverman game is a two person zero sum game defined in terms of two sets, S_1 and S_2, of positive numbers and two parameters, the threshold $T > 1$ and the penalty $\nu > 0$. Players I and II choose numbers independently from S_1 and S_2, respectively. The higher number wins 1, unless it is at least T times as large as the other, in which case it loses ν. If the numbers are equal the payoff is zero.

Such a game might be thought of as an imperfect model for various bidding or spending situations in which within some bounds the higher bidder or the bigger spender "wins", but loses if it is overdone. Some situations which come to mind are spending on armaments, advertising spending, or sealed bids in an auction.

Most previous work on such games has dealt either with symmetric games, where $S_1 = S_2$, or with disjoint games, where $S_1 \cap S_2 = \phi$. A version of the symmetric game on a special discrete set S is described in [3, p. 212]. In [1], Evans examined the symmetric game on (a,b), where $0 < a < b \leq \infty$, obtained necessary and sufficient conditions that an optimal strategy exist and gave an optimal strategy in the case where one exists. Symmetric games on an arbitrary discrete set S are solved in [2] for all T and ν except for ν too near zero in some cases. An analogous game with $S = [a,\infty)$, $a > 0$, with payoff a certain continuous function of y/x, is examined in [5].

Nonsymmetric Silverman games were first considered by Heuer in [4], where the game with S_1 the set of positive odd integers and S_2 the evens was solved for all T and ν. This work was extended to arbitrary discrete and disjoint S_1 and S_2 in [7], where a classification into 8 classes and solutions are obtained for $\nu \geq 1$ and all T, and partial results are obtained for $\nu < 1$.

Nearly all the games studied in the above-mentioned papers have optimal strategies whose support is a bounded subset of the corresponding strategy set, and thus in the discrete case optimal strategies are of finite type. The reason for this, at least when $\nu \geq 1$, is made clear in [6], where it is shown that Silverman games with penalty ≥ 1 are part of a much larger class of games which always have bounded optimal strategies.

In this work we begin to analyze the vast class of discrete Silverman games that lie between the extremes of $S_1 = S_2$ and $S_1 \cap S_2 = \phi$. We recall a few facts about these two extreme cases. When $S_1 = S_2$ the game always reduces to a (2n+1) by (2n+1) game for some $n \geq 0$. The <u>essential subgame</u> is the game on the essential set

$$W = \{e_1, e_2, \ldots, e_{n+1}, f_1, \ldots f_n\},$$

where $e_{n+1} = \langle Te_1 \rangle = \langle Tc_1 \rangle$ (here $\langle x \rangle$ denotes the largest element of S less than x, and c_1 is the smallest element of S_1), and $f_i = \langle Te_{i+1} \rangle$. Further, $f_i = \langle Tc \rangle$ whenever $e_i < c \leq e_{i+1}$.

In the disjoint case [7] there are 8 classes. In classes 1A, 2A and 2B, at least one player has an optimal pure strategy, and when $\nu \geq 1$ both do, so the game has a saddle point.

In classes 3A and 3B the game reduces to 2 by 2, and in the remaining classes, 4A.k, 4B.k and 5A.k, the game reduces to (2k+1) by (2k+1).

In the work that follows we begin a systematic analysis of Silverman games where S_1 and S_2 are arbitrary discrete sets of positive numbers and the penalty is ≥ 1. There are always finite subsets W_1 of

S_1 and W_2 of S_2 such that optimal strategies for the subgame on $W_1 \times W_2$ are optimal for the full game on $S_1 \times S_2$, and a principal objective is to find minimal subsets with this property.

In Section 5 we define balanced Silverman games, and thereafter limit our study to these games. We show in Sections 8 to 11 how all balanced Silverman games reduce to nine fundamental types, one of which is 2 by 2, four of which are larger games of even order, and four of which are of odd order. We think these are all irreducible, and discuss the evidence for this in Section 13.

2. Games with saddle points.

The theorems in [7] dealing with classes 1A, 2A and 2B do not depend on the strategy sets being disjoint, and include all Silverman games where at least one player has an optimal pure strategy, except the symmetric 1 by 1 case:

THEOREM 2.1. In the symmetric Silverman game (S,T,ν), suppose that there is an element c in S such that $c < Tc_i$ for all c_i in S, and that $S \cap (c,Tc) = \phi$. Then pure strategy c is optimal.

PROOF. Let $A(x,y)$ be the payoff function. By symmetry the game value is 0. Since $A(c,y) = 1,0$ or ν according as $y < c$, $y = c$ or $y \geq Tc$, we have $A(c,y) \geq 0$ for every y in S. □

In this theorem, as in those referred to in the preceding paragraph, no assumption of discreteness is made.

3. <u>The</u> 2 <u>by</u> 2 <u>games</u>.

For the remainder of the paper we assume that S_1 and S_2 are discrete. It turns out that a great many discrete Silverman games are reducible to 2 by 2 games, in the sense that each player has a 2-component optimal mixed strategy. In this section we shall identify all irreducible 2 by 2 Silverman games, and in the next section are some theorems giving conditions under which games reduce to 2 by 2. "Game" hereafter will always mean "Silverman game."

It is clear from the payoff rule for Silverman games that if the elements in each S_i are listed in increasing order, the entries in each row of the payoff matrix are subject to the order $-\nu$, 1, 0, -1, ν, and columns, from top to bottom, the opposite order. It is easy to see that a 2 by 2 game with ν or $-\nu$ on the diagonal reduces by dominance to a 1 by 1 game. (In Section 5 we shall see that a game of any size having $|A(i,i)| = \nu$ for some i is reducible by dominance.) Since interchanging S_1 and S_2 replaces the game matrix A by its negative transpose, which we shall denote by A', it will suffice to find all irreducible 2 by 2 game matrices where the first nonzero diagonal element is -1.

Subject to the above restriction, and taking into account the row and column order and dominance considerations, one finds that there are just 3 possible first rows, namely

$0\ \ -1, \quad 0\ \ \nu, \quad \text{and} \quad -1\ \ \nu.$

It is straightforward then to verify that there are exactly 8 irreducible 2 by 2 game matrices, namely the following four and their duals (negative transposes):

(A) (T = 4)

	2	4
1	-1	ν
3	1	-1

(B) (T = 3)

	2	3
1	-1	ν
2	0	-1

(C) (T = 3)

	1	3
1	0	ν
2	1	-1

(D) (T = 3)

	2	3
1	-1	ν
3	1	0

(The first row 0 -1 occurs only in (C').)

The unique optimal mixed strategy $P = (p_1, p_2)$ for the row player, $Q = (q_1, q_2)$ for the column player, and the game value V are given below for convenience.

(A) $P = (2,\nu+1)/(\nu+3)$, $Q = (\nu+1,2)/(\nu+3)$, $V = (\nu-1)/(\nu+3)$

(B) $P = (1,\nu+1)/(\nu+2)$, $Q = (\nu+1,1)/(\nu+2)$, $V = -1/(\nu+2)$

(C) $P = (2,\nu)/(\nu+2)$, $Q = (\nu+1,1)/(\nu+2)$, $V = (\nu/(\nu+2)$

(D) $P = (1,\nu+1)/(\nu+2)$, $Q = (\nu,2)/(\nu+2)$, $V = \nu/(\nu+2)$.

4. Some games which reduce to 2 by 2 when $\nu \geq 1$.

The game of case (A) above and its dual (A') are the reduced games of Classes 3A and 3B in the disjoint case [7]. However, many games where $S_1 \cap S_2 \neq \phi$ also reduce to these 2 by 2 games, as we see in the first two theorems below. From now on we assume also that $\nu \geq 1$.

Let $S_1 = \{c_1, c_2, c_3, \ldots\}$, $S_2 = \{d_1, d_2, d_3, \ldots\}$, with $c_i < c_{i+1}$ and $d_i < d_{i+1}$ for each i. We assume without loss of generality that $1 = c_1 \leq d_1$. Extending slightly a notation used in [2],

(4.0.1) $\langle x \rangle_i$ denotes the largest element of S_i less than x.

When the context makes clear which S_i is involved we may simply write $\langle x \rangle$. E.g., in the equation $d_k = \langle c_j T \rangle$ it is understood that d_k is in S_2. For each i, let

$$(4.0.2) \qquad \begin{cases} c_i^* = \min\, [d_i, \infty) \cap S_1, \text{ if } [d_i, \infty) \cap S_1 \neq \phi \\ d_i^* = \min\, [c_i, \infty) \cap S_2, \text{ if } [c_i, \infty) \cap S_2 \neq \phi. \end{cases}$$

For given S_1, S_2 and T, define integers m and r by

$$(4.0.3) \qquad \begin{cases} c_m = \langle d_1 T \rangle; \\ d_r = \langle c_1 T \rangle = \langle T \rangle. \end{cases}$$

Let $P = (p_1, p_2, p_3, \ldots)$ and $Q = (q_1, q_2, q_3, \ldots)$ denote the mixed strategies, on S_1 and S_2 respectively, which assign probabilities p_i to c_i and q_i to d_i for $i = 1,2,3,\ldots$. The payoff for (x,y) in $S_1 \times S_2$ is always denoted by $A(x,y)$. The expected payoff for mixed strategies γ,δ is denoted by $E(\gamma,\delta)$.

Consider the game with $S_1 = \{1,3,5,7,9,29,42,66\}$, $S_2 = \{2,4,6,7,28,36,66,89\}$ and $T = 10$. Here $c_m = 9$, $d_r = 7$, and the subgame on $\{1,9\} \times \{2,28\}$ has the matrix of case (A) in Section 3. Optimal strategies for this 2 by 2 game are $P = (2,\nu+1)/(\nu+3)$, $Q = (\nu+1,2)/(\nu+3)$, and the game value is $V = (\nu-1)/(\nu+3)$. Although there are no dominated strategies in S_1 or S_2 (see game matrix below), we shall see that P and Q are optimal for the full game on $S_1 \times S_2$. We partition the matrix as follows:

		$(\nu+1)$				(2)			
		2	4	6	7	28	36	66	89
(2)	1	-1	-1	-1	-1	ν	ν	ν	ν
	3	1	-1	-1	-1	-1	ν	ν	ν
	5	1	1	-1	-1	-1	-1	ν	ν
	7	1	1	1	0	-1	-1	-1	ν
$(\nu+1)$	9	1	1	1	1	-1	-1	-1	-1
	29	$-\nu$	1	1	1	1	-1	-1	-1
	42	$-\nu$	$-\nu$	1	1	1	1	-1	-1
	66	$-\nu$	$-\nu$	$-\nu$	1	1	1	0	-1

Against $\{2,28\}$, the strategies 3,5,7,9 in S_1 are equivalent, as are 29,42,66, and the latter group has expectation less than V. Against $\{1,9\}$ the strategies 2,4,6,7 in S_2 are equivalent, as are 28,36,66,89.

Consequently, strategies optimal on the 2 by 2 subgame are optimal for the full game. Theorem 4.1 below gives general conditions under which such a reduction to a case (A) 2 by 2 game is possible. In the notation of that theorem and of (4.0.2) we have j = 1, $c_1^* = 3$, $d_k = 28$ and $c_k^* = 29 \geq d_1T$ in the above example.

THEOREM 4.1. Assume that

(4.1.1) $d_r < c_m$ (i.e., that $S_2 \cap [c_m, T) = \phi$);

(4.1.2) $\exists\, d_j < c_m$ such that if $d_k = \langle c_j^* T\rangle$ then

$S_1 \cap [d_k, d_jT) = \phi$.

(Note that then $d_j > 1$. See remark below.)

Then the game value is $(\nu-1)/(\nu+3)$, and the following strategies γ and δ are optimal:

$$\begin{array}{ccccc} \underline{\gamma} & & \underline{\delta} & & \\ p_1 & = & q_k & = & 2/(\nu+3) \\ p_m & = & q_j & = & (\nu+1)/(\nu+3) \end{array}$$

REMARK. If $d_j = 1$, then $c_j^* = 1$, $d_k = \langle T\rangle$, so $d_k < c_m$ by (4.1.1). Then $c_k^* \leq c_m < d_1T$, in contradiction to (4.1.2). Thus $d_j > 1$.

PROOF of theorem. Let $V = (\nu-1)/(\nu+3)$. We show first that $E(\gamma,d) \geq V$ for all d in S_2. If $d < c_m$, then $d < c_m < d_1T \leq dT$, so $A(c_m,d) = 1$. Also, $c_1 \leq c_m < dT$, so $A(c_1,d) \geq -1$. Thus $E(\gamma,d) \geq p_m - p_1 = V$.

If $d \ge c_m$, then $d \ge T = c_1T$, so $A(c_1,d) = \nu$. Also, $A(c_m,d) \ge -1$, so $E(\gamma,d) \ge \nu p_1 - p_m = V$.

Next we show that $E(c,\delta) \le V$ for all c in S_1. If $c < d_j$, then $c < d_j < c_m \le Tc$, so $A(c,d_j) = -1$. Since $A(c,d_k) \le \nu$, we have $E(c,\delta) \le -q_j + \nu q_k = V$.

If $d_j \le c < d_k$, then $c_j^* \le c$, so $c < d_k < c_j^*T \le Tc$, and therefore $A(c,d_k) = -1$. Moreover, $d_j \le c \Rightarrow A(c,d_j) \le 1$, so we have $E(c,\delta) \le q_j - q_k = V$.

Finally, if $c \ge d_k$, then $c \ge c_k^* \ge d_jT$ by (4.1.2), so $A(c,d_j) = -\nu$. But $A(c,d_k) \le 1$, so $E(c,\delta) \le -\nu q_j + q_k$ $= -(\nu^2+\nu-2)/(\nu+3) \le 0 \le V$. □

THEOREM 4.2. Assume that

(4.2.1) $c_m < d_r$ (i.e., that $S_1 \cap [d_r,d_1T) = \phi$);

(4.2.2) $\exists\, c_j < d_r$ such that if $c_k = \langle d_j^*T \rangle$ then $S_2 \cap [c_k,c_jT) = \phi$.

(Note that then $1 < c_j < c_k$. See remark below.)

Then the game value is $(-\nu+1)/(\nu+3)$, and the following strategies, γ and δ, are optimal:

$\underline{\gamma}$		$\underline{\delta}$
p_j	=	$q_r = (\nu+1)/(\nu+3)$
p_k	=	$q_1 = 2/(\nu+3)$.

REMARK. If $j = 1$, then $c_k < d_1T$, and (4.2.1) then implies that $c_k < T$, and therefore $c_k \le c_m$. Then

(4.2.1) further implies that $d_k^* \le d_r < T$. But $d_k^* \ge c_k$, so that (4.2.2) implies $d_k^* \ge c_jT$, a contradiction. Thus $j > 1$. Furthermore, from (4.2.2) we have $c_j < d_r < T < c_jT$, but $S_2 \cap [c_k, c_jT) = \phi$. Therefore $c_k > c_j$.

PROOF of theorem. Let $V = (-\nu+1)/(\nu+3)$. We show first that $E(\gamma,d) \ge V$ for all d in S_2. (i) If $d < c_j$ then $d < c_j < d_r < T \le dT$, so $A(c_j,d) = 1$. Also, $A(c_k,d) \ge -\nu$, so $E(\gamma,d) \ge p_j - \nu p_k = V$.
(ii) If $c_j \le d < c_k$, then $d_j^* \le d$, so $d < c_k < d_j^*T \le dT$, and $A(c_k,d) = 1$. Also, $c_j \le d \Rightarrow A(c_j,d) \ge -1$, so $E(\gamma,d) \ge -p_j + p_k = V$. (iii) If $d \ge c_k$, then $d \ge c_jT$ by (4.2.2), so $A(c_j,d) = \nu$. Since $A(c_k,d) \ge -1$, we have $E(\gamma,d) \ge \nu p_j - p_k = (\nu^2+\nu-2)/(\nu+3) \ge 0 \ge V$.

We complete the proof by showing that $E(c,\delta) \le V$ for all c in S_1. (i) If $c < d_r$, then $c < d_r < T \le cT$, so $A(c,d_r) = -1$. Also, $d_1 \le d_r < cT$, so $A(c,d_1) \le 1$. Thus $E(c,\delta) \le q_1 - q_r = V$. (ii) If $c \ge d_r$, then by (4.2.1) we have $c \ge d_1T$, so $A(c,d_1) = -\nu$. Since $A(c,d_r) \le 1$, we have $E(c,\delta) \le -\nu q_1 + q_r = V$. □

The next two theorems give conditions under which the game reduces to the 2 by 2 game of case (B) or its dual (B'). Examples illustrating Theorems 4.3, 4.5 and 4.7 are given following Theorem 4.7.

THEOREM 4.3. Assume that

(4.3.1) $$c_m = d_r,$$

(4.3.2) $$c_{m+1} \geq d_r T, \quad \text{and}$$

(4.3.3) $$\exists\, c_i < d_r \text{ such that } c_i T \leq d_{r+1} < c_m T.$$

Then $V = -1/(\nu+2)$, and the following strategies, γ and δ, are optimal:

$\underline{\gamma}$		$\underline{\delta}$		
p_m	$=$	q_r	$=$	$(\nu+1)/(\nu+2)$
p_i	$=$	q_{r+1}	$=$	$1/(\nu+2)$.

PROOF. We show first that $E(\gamma,d) \geq -1/(\nu+2)$ for all d in S_2. (i) If $d \leq c_m$, then $A(c_m,d) \geq 0$ because $d \leq c_m < dT$. Since $c_i < d_r < T \leq dT$, $A(c_i,d) \geq -1$. Thus $E(\gamma,d) \geq -p_i = -1/(\nu+2)$. (ii) If $d > c_m$, then $d \geq d_{r+1} \geq c_i T$, so $A(c_i,d) = \nu$. We also have $A(c_m,d) \geq -1$, so $E(\gamma,d) \geq \nu p_i - p_m = -1/(\nu+2)$.

We complete the proof by showing that $E(c,\delta) \leq -1/(\nu+2)$ for all c in S_1. (i) If $c < d_r$, then $c < d_r < cT$ so $A(c,d_r) = -1$. Since $A(c,d_{r+1}) \leq \nu$, we have $E(c,\delta) \leq -q_r + \nu q_{r+1} = -1/(\nu+1)$. (ii) If $c = d_r$, then $A(c,d_r) = 0$, and since $c = d_r < d_{r+1} < c_m T = cT$, we have $A(c,d_{r+1}) = -1$. Thus $E(c,\delta) = -q_{r+1} = -1/(\nu+2)$. (iii) If $c > d_r$, then $c \geq c_{m+1} \geq d_r T$, so $A(c,d_r) = -\nu$. Also, $c \geq d_r T = c_m T > d_{r+1}$, so $A(c,d_{r+1}) \leq 1$. Thus $E(c,\delta) \leq -\nu q_r + q_{r+1} = (-\nu^2-\nu+1)/(\nu+2) \leq -1/(\nu+2)$. □

Similarly, one proves the dual:

THEOREM 4.4. Assume that

(4.4.1) $c_m = d_r$,

(4.4.2) $d_{r+1} \geq c_m T$, and

(4.4.3) $\exists\, d_i < c_m$ such that $d_i T \leq c_{m+1} < d_r T$.

Then $V = 1/(\nu+2)$, and the following strategies are optimal:

$$p_m = q_r = (\nu+1)/(\nu+2)$$

$$p_{m+1} = q_i = 1/(\nu+2).$$

The next theorem gives conditions under which the game reduces to a type (C) 2 by 2.

THEOREM 4.5. Assume that

(4.5.1) $c_{m-1} = d_r$,

(4.5.2) $c_m < d_{r+1} < c_m T$, and

(4.5.3) $c_{m-1} T \leq d_{r+1} \leq c_{m+1}$.

Then the game value is $\nu/(\nu+2)$, and the following strategies, γ and δ, are optimal:

$$\gamma: \quad p_{m-1} = 2/(\nu+2) \quad , \qquad p_m = \nu/(\nu+2)$$

$$\delta: \quad q_r = (\nu+1)/(\nu+2), \qquad q_{r+1} = 1/(\nu+2).$$

PROOF. Let $V = \nu/(\nu+2)$. We show first that $E(\gamma,d) \geq V$ for all d in S_2. (i) If $d \leq d_r$, then $d \leq c_{m-1} < dT$, so $A(c_{m-1},d)$ is 1 or 0. Since $d < c_m < dT$, we have $A(c_m,d) = 1$. Thus $E(\gamma,d) \geq p_m = V$. (ii) If

$d \geq d_{r+1}$, then by (4.5.3), $A(c_{m-1},d) = \nu$. Since by (4.5.2), $d > c_m$, we have $A(c_m,d) \geq -1$. Thus $E(\gamma,d) \geq \nu p_{m-1} - p_m = V$.

We complete the proof by showing that $E(c,\delta) \leq V$ for all c in S_1. (i) If $c \leq c_{m-1}$, then $c \leq d_r < cT$, so $A(c,d_r)$ is 0 or -1. Hence $E(c,\delta) \leq 0q_r + \nu q_{r+1} = V$. (ii) If $c = c_m$, then $d_r = c_{m-1} < c_m <-d_rT$, so $A(c_m,d_r) = 1$. From (4.5.2) we have $A(c_m,d_{r+1}) = -1$, so $E(c_m,\delta) = q_r - q_{r+1} = V$. (iii) If $c \geq c_{m+1}$, then $c \geq d_rT$ by (4.5.1) and (4.5.3), so that $A(c,d_r) = -\nu$, and by (4.5.3), $A(c,d_{r+1}) \leq 1$. Thus $E(c,\delta) \leq -\nu q_r + q_{r+1} = (-\nu^2-\nu+1)/(\nu+2) \leq 0 < V$. □

The dual theorem is the following.

THEOREM 4.6. Assume that

(4.6.1) $d_{r-1} = c_m$,

(4.6.2) $c_{m+1} < d_rT$, and

(4.6.3) $d_{r-1}T \leq c_{m+1} \leq d_{r+1}$.

(Note that now $d_r < c_1T \leq d_1T \leq c_{m+1} \Rightarrow d_r < c_{m+1}$.)

Then the game value is $V = -\nu/(\nu+2)$, and the following strategies are optimal.

$$\gamma: \quad p_m = (\nu+1)/(\nu+2), \quad p_{m+1} = 1/(\nu+2)$$

$$\delta: \quad q_{r-1} = 2/(\nu+2), \quad q_r = \nu/(\nu+2).$$

The proof is similar to that of Theorem 4.5. □

The next theorem deals with games that reduce to 2 by 2 games of type (D).

THEOREM 4.7. Assume that

(4.7.1) $T > d_1 \notin S_1$,

(4.7.2) $T \le c_r = d_k < d_1T$ and

(4.7.3) $d_{k+1} \ge d_kT$.

Then the game value is $V = \nu/(\nu+2)$, and the following strategies, γ and δ, are optimal:

$$\gamma:\quad p_1 = 1/(\nu+2), \qquad p_r = (\nu+1)/(\nu+2)$$

$$\delta:\quad q_1 = \nu/(\nu+2), \quad q_k = 2/(\nu+2)$$

PROOF. We show first that $E(\gamma,d) \ge V$ for all d in S_2. (i) If $d < c_r$, then since $c_r < d_1T \le dT$ we have $A(c_r,d) = 1$. By (4.7.1), $d > 1$, so $A(1,d) \ge -1$. Thus $E(\gamma,d) \ge -p_1 + p_r = V$. (ii) If $d = c_r = d_k$, then by (4.7.2), we have $A(1,d_k) = \nu$, and $A(c_r,d_k) = 0$, so $E(\gamma,d) = \nu p_1 = V$. (iii) If $d > d_k$, then by (4.7.3), $A(c_r,d) = \nu = A(c_1,d)$, so that $E(\gamma,d) = \nu > V$.

We complete the proof by showing that $E(c,\delta) \le V$ for all c in S_1. (i) If $c \le d_1$, then by (4.7.1) we have $A(c,d_1) = -1$. Since $A(c,d_k) \le \nu$, $E(c,\delta) \le -q_1 + \nu q_k = V$. (ii) If $d_1 < c \le d_k$, then (4.7.2) implies $d_1 < c < d_1T$, so $A(c,d_1) = 1$. Also, $c \le d_k < d_1T < cT$, which implies that $A(c,d_k)$ is 0 or -1.

Thus $E(c,\delta) \le q_1.1 + q_2.0 = V$. (iii) If $c > -d_k$, then $c \ge c_{r+1} \ge d_1T$, so $A(c,d_1) = -\nu$. Since $A(c,d_k) \le 1$, we have $E(c,\delta) \le -\nu q_1 + q_k = (-\nu^2+2)/(\nu+2) \le \nu/(\nu+2) = V$. □

The dual case, (D'), does not occur under the convention that $c_1 \le d_1$. Section 10 shows how another large class of games reduces to 2 by 2 games of type A or A'.

Below we give examples of games which reduce to 2 by 2 games of types B, C and D as indicated by Theorems 4.3, 4.5 and 4.7. The asterisks in the margin indicate the active strategies, and the separating lines aid in seeing that the optimal mixed strategies for the 2 by 2 subgame are optimal for the full game.

Type B,
Theorem 4.3.
$T = 10$

					*	*		
		1	3	4	9	40	50	95
	1	0	-1	-1	-1	ν	ν	ν
	2	1	-1	-1	-1	ν	ν	ν
*	3	1	0	-1	-1	ν	ν	ν
*	9	1	1	1	0	-1	-1	ν
	90	$-\nu$	$-\nu$	$-\nu$	$-\nu$	1	1	-1
	96	$-\nu$	$-\nu$	$-\nu$	$-\nu$	1	1	1

Type C,
Theorem 4.5.
$T = 10$

					*	*			
		1	3	4	5	55	65	80	85
	1	0	-1	-1	-1	ν	ν	ν	ν
	2	1	-1	-1	-1	ν	ν	ν	ν
	3	1	0	-1	-1	ν	ν	ν	ν
*	5	1	1	1	0	ν	ν	ν	ν
*	9	1	1	1	1	-1	-1	-1	-1
	60	$-\nu$	$-\nu$	$-\nu$	$-\nu$	1	-1	-1	-1
	70	$-\nu$	$-\nu$	$-\nu$	$-\nu$	1	1	-1	-1
	85	$-\nu$	$-\nu$	$-\nu$	$-\nu$	1	1	1	0

Type D,
Theorem 4.7.
T = 10

		*			*		
		2	3	5	12	120	130
*	1	-1	-1	-1	ν	ν	ν
	3	1	0	-1	-1	ν	ν
	4	1	1	-1	-1	ν	ν
*	12	1	1	1	0	ν	ν
	60	$-\nu$	$-\nu$	$-\nu$	1	-1	-1
	90	$-\nu$	$-\nu$	$-\nu$	1	-1	-1

5. Reduction by dominance.

In [6], it is shown that every discrete Silverman game with $\nu \geq 1$ reduces by dominance to a finite game, and in [7], it is shown that if $S_i \cap [a,b] = \phi$, where a and b are elements of S_{3-i}, then b is dominated by a. In this section we shall discuss four types of dominance for Silverman games, including the above two. Through repeated reduction of the strategy sets S_1 and S_2 by means of these four types of dominance we obtain what we call pre-essential sets $\tilde{W}_1 \subset S_1$ and $\tilde{W}_2 \subset S_2$. These are minimal subsets in the sense that no further reduction is possible through the use of these four types of dominance.

In the symmetric case, where $S_1 = S_2$, the common reduced set at this stage is the essential set of Evans and Heuer [2]. In the general case, $\tilde{W}_1$ and $\tilde{W}_2$ need not yet be essential sets, in the sense that optimal strategies for the game on $\tilde{W}_1 \times \tilde{W}_2$ must assign positive probabilities to each of their elements. In Sections 8 to 11 we discuss conditions under which further reduction is possible, and obtain, for what we call balanced Silverman games, what appear to be irreducible subgames with the property that optimal strategies for the subgame are optimal for the full game.

We have the following four types of dominance.

A. The reduction to finite sets.

In [6] it has been shown that if $d_j \geq Tc_m$ then d_1 dominates d_j, and any $c_i \geq Td_r$ is dominated by c_1. For the convenience of the reader we give a brief sketch here. If $d_j \geq Tc_m$ then $A(c_i,d_j) = \nu$ for all $i \leq m$, and therefore $A(c_i,d_1) \leq A(c_i,d_j)$ because all $A(x,y) \leq \nu$. For $i > m$, then (by definition of m) $c_i \geq Td_1$ so that $A(c_i,d_1) = -\nu \leq A(c_i,d_j)$ because all $A(x,y) \geq -\nu$. The argument for $c_i \geq Td_r$ is similar. The following table makes the argument graphically:

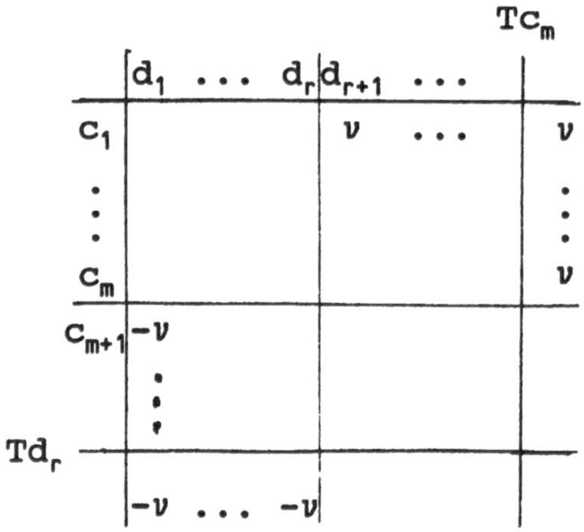

Tc_m
d_1 ... d_r d_{r+1} ...
c_1
ν ... ν
c_m
ν
c_{m+1} $-\nu$
Td_r
$-\nu$ ... $-\nu$

Thus we reduce our strategy sets to $S_1 \cap (0,Td_r)$ and $S_2 \cap (0,Tc_m)$.

B. Two elements of S_{3-i} in an S_i-interval.

As shown in [7], if $c_k < d_j < d_{j+1} < c_{k+1}$, then d_j dominates d_{j+1}, and we delete d_{j+1} from S_2. Similarly, if $d_j < c_k < c_{k+1} < d_{j+1}$ or $c_k < c_{k+1} < d_1$, then c_k dominates c_{k+1} and we eliminate c_{k+1}. Also, if S_{3-i} has two or more elements greater than the largest element of S_i, the first of these greater elements dominates the others. The argument is illustrated in the following table for the case of two elements of S_1 between consecutive elements of S_2: (T=10)

	...	4	5	6	...	40	55	...	440	460	510
⋮											
45		$-\nu$	1	1	...	1	-1	...	-1	ν	ν
51		$-\nu$	$-\nu$	1	...	1	-1	...	-1	-1	ν

45 dominates 51 in S_1.

C. Two elements of S_{3-i} in a TS_i-interval.

LEMMA 5.1. Assume that S_1 and S_2 have been truncated as described in A.

(a) If for some $k < m$, we have

(5.1.1) $Tc_k \le d_j < d_{j+1} < Tc_{k+1}$, then d_{j+1} dominates d_j.

(b) If for some $k < r$, we have

(5.1.2) $Td_k \le c_j < c_{j+1} < Td_{k+1}$, then c_{j+1} dominates c_j.

Before giving a formal proof we illustrate the argument for part (b) in the following table: (T=10)

	4	5	...	43	45	48	50	...	399
43	$-\nu$	1	...	0	-1	-1	-1	...	-1
48	$-\nu$	1	...	1	1	0	-1	...	-1

48 dominates 43 in S_1

PROOF. (a) If $c \le c_k$ then $A(c,d_j) = A(c,d_{j+1}) = \nu$. If $c_{k+1} \le c < d_j$, then $A(c,d_j) = A(c,d_{j+1}) = -1$. If $c = d_j$, then $A(c,d_j) = 0 > -1 = A(c,d_{j+1})$. If $d_j < c < d_{j+1}$, then $A(c,d_j) = 1 > -1 = A(c,d_{j+1})$. Since S_1 has been truncated at Td_r, and by (5.1.1) $d_j \ge Tc_1 > d_r$, S_1 has no elements $\ge Td_j$. If $c = d_{j+1}$ then $d_j < c < Td_j$ so $A(c,d_j) = 1$ while $A(c,d_{j+1}) = 0$. If $d_{j+1} < c < Td_j$, then $A(c,d_j) = A(c,d_{j+1}) = 1$, so we have $A(c,d_{j+1}) \le A(c,d_j)$ for all c in S_1.

(b) The proof here is similar. □

D. Two elements of TS_{3-i} in an S_i-interval.

LEMMA 5.2. (a) Suppose that for some $d_j < d_r$ we have $\langle Td_j \rangle_1 = \langle Td_{j+1} \rangle_1 = c_k$; i.e.,

(5.2.1) $\quad c_k < Td_j < Td_{j+1} \le c_{k+1}$.

Then d_{j+1} dominates d_j.

(b) If for some $c_j < c_m$ we have $\langle Tc_j \rangle_2 = \langle Tc_{j+1} \rangle_2 = d_k$; i.e.,

(5.2.2) $\quad d_k < Tc_j < Tc_{j+1} \le d_{k+1}$,

then c_{j+1} dominates c_j.

Before giving the proof we illustrate the argument for part (b) in the following table: (T=10)

	3	4	5	6	7	...	38	60	...	399
4	1	0	-1	-1	-1	...	-1	ν	...	ν
6	1	1	1	0	-1	...	-1	ν	...	ν

6 dominates 4 in S_1.

PROOF. (a) If $c < d_j$ then $c < d_j < d_{j+1} \le d_r < cT$, so $A(c,d_j) = A(c,d_{j+1}) = -1$. If $c = d_j$ then $A(c,d_j) = 0 > -1 = A(c,d_{j+1})$. If $d_j < c < d_{j+1}$ then $A(c,d_j) = 1 > -1 = A(c,d_{j+1})$. If $c = d_{j+1}$, then $A(c,d_j) = 1 > 0 = A(c,d_{j+1})$. If $d_{j+1} < c \le c_k$, then $A(c,d_j) = A(c,d_{j+1}) = 1$. If $c \ge c_{k+1}$ then $A(c,d_j) = A(c,d_{j+1}) = -\nu$. In all cases we have $A(c,d_{j+1}) \le A(c,d_j)$.

(b) The proof here is similar. □

By "step A" applied to a given pair of strategy sets S_1 and S_2 we shall mean the removal of all dominated elements of the type discussed in (A) above. Similar understandings apply to "step B," "step C" and "step D." These steps may be further broken down into A_1, A_2, B_1, B_2, etc., where step A_1 refers to removal from S_1 of dominated elements of type A, etc. It is convenient to assume that after each of the steps B_i, C_i, D_i the elements of S_i are renumbered so that the k-th element in increasing order has subscript k again.

It is sometimes the case that after steps A, B, C and D have been taken, further reduction is possible by repeating these steps. However, since after step A the strategy sets are finite (we are assuming the original strategy sets to be discrete), after some finite number of the above steps no further reduction in this way is possible.

Let $\tilde{W}_1$ and $\tilde{W}_2$ be the subsets of S_1 and S_2 that remain when the cycle A_1, A_2, B_1, B_2, C_1, C_2, D_1, D_2, has been repeated until no further reduction occurs. We shall call $\tilde{W}_1$ and $\tilde{W}_2$ the <u>pre-essential strategy sets</u>, and write e_j and f_j for the j-th element of $\tilde{W}_1$, $\tilde{W}_2$ respectively, in increasing order. The notation $\langle e_i T\rangle_2$ in this context refers to the largest element of $\tilde{W}_2$ smaller than $e_i T$; similarly, $\langle f_i T\rangle_1$ is the largest element of $\tilde{W}_1$ smaller than $f_i T$. Many of these games are further reducible, in the sense that there are proper subsets W_i of $\tilde{W}_i$ such that optimal strategies for the game on $W_1 \times W_2$ are optimal for the game on $\tilde{W}_1 \times \tilde{W}_2$, and therefore also for the full original game. For what we shall call balanced games, this reduction is treated in Sections 8-11. We shall refer to the game on $\tilde{W}_1 \times \tilde{W}_2$ as the <u>semi-reduced</u> game.

(5.2.3) Let n and s be the integers such that

$$e_{n+1} = \langle f_1 T\rangle \text{ and } f_{s+1} = \langle e_1 T\rangle.$$

THEOREM 5.3. $|\hat{W}_1| = |\hat{W}_2| = n+s+1$, and for $k = 1,\ldots,s+1$, $e_{n+k} = \langle f_k T\rangle$. For $k = 1,\ldots,n+1$, $f_{s+k} = \langle e_k T\rangle$. Thus

$$\hat{W}_1 = \{e_1, e_2, \ldots, e_{n+1}, \langle f_2 T\rangle, \ldots, \langle f_{s+1} T\rangle\} \text{ and}$$

$$\hat{W}_2 = \{f_1, f_2, \ldots, f_{s+1}, \langle e_2 T\rangle, \ldots, \langle e_{n+1} T\rangle\}.$$

PROOF. $\hat{W}_1$ has no element larger than $f_{s+1}T$ and $\hat{W}_2$ none larger than $e_{n+1}T$ because of invariance under step A. $\hat{W}_1$ has no more than s+1 elements $\geq e_{n+1}$, for otherwise we would have

$$Tf_k \leq e_j < e_{j+1} < Tf_{k+1}$$

for some $k < s+1$ and some $j > n+1$, contrary to invariance under step C. The s+1 elements $e_{n+1} = \langle f_1 T\rangle$, $\langle f_2 T\rangle, \ldots, \langle f_{s+1} T\rangle$ must be distinct because of invariance under step D. Thus $\hat{W}_1$ has exactly n+s+1 elements, with $e_{n+k} = \langle f_k T\rangle$ for $k = 1,\ldots,s+1$. A dual argument shows the corresponding facts for $\hat{W}_2$. □

The following examples illustrate.

EXAMPLE 5.4. Let $S_1 = \{1,2,3,5,7,8,11,20,25,31,41,48,55,70,75,81,88,95,100,\ldots\}$, $S_2 = \{1,4,5,6,8,9,15,29,30,38,49,58,65,75,80,89,98,105,\ldots\}$ and $T = 10$. Step A removes all elements ≥ 90 from S_1 and all elements ≥ 80 from S_2. Step B removes 3, 25, 48, 88 from S_1 and 30, 65 from S_2. Step C removes 11, 20, 70 from S_1 and 29, 38 from S_2. Step D changes nothing, and the reduced sets after this first pass are $S_1' = \{1,2,5,7,8,31,41,55,75,81\}$, $S_2' = \{1,4,5,6,8,9,15,49,58,75\}$. In the second pass, step A changes nothing, step B removes 41 from S_1 and 15 from S_2. Step C changes nothing and step D removes 1 from S_1 and 4 from S_2. A third pass leaves the sets unchanged, and the pre-essential sets are

$$\tilde{W}_1 = \{2,5,7,8,31,55,75,81\}$$

$$\tilde{W}_2 = \{1,5,6,8,9,49,58,75\} .$$

Here $n = 3$, $s = 4$, and each set has $n+s+1 = 8$ elements.

EXAMPLE 5.5. Let $S_1 = \{1,2,4,5,7,8,9,20,28,36,50,59,85,95,101,\ldots\}$, $S_2 = \{1,3,4,5,6,8,9,15,28,35,52,59,84,95,105,\ldots\}$, $T = 10$. After one pass of steps A, B, C, D we have the pre-essential sets

$$\tilde{W}_1 = \{1,2,5,8,9,28,36,59,85\}$$

$$\tilde{W}_2 = \{1,3,5,8,9,15,35,59,84\} ,$$

with $n = s = 4$, and each reduced set has $2n+1 = 9$ elements.

Following are the payoff matrices for the reduced games in these two examples. In accordance with our convention that Player I has the smallest pure strategy, we interchange $\tilde{W}_1$ and $\tilde{W}_2$ in the first, making n = 4, s = 3. In general the matrix has n subdiagonals with each element being -1 or 0, an s by s triangle of $-\nu$s in the lower left corner, s superdiagonals of 1s or 0s and an n by n triangle of νs in the upper right corner.

Example 5.4

	2	5	7	8	31	55	75	81
1	-1	-1	-1	-1	ν	ν	ν	ν
5	1	0	-1	-1	-1	ν	ν	ν
6	1	1	-1	-1	-1	-1	ν	ν
8	1	1	1	0	-1	-1	-1	ν
9	1	1	1	1	-1	-1	-1	-1
49	$-\nu$	1	1	1	1	-1	-1	-1
58	$-\nu$	$-\nu$	1	1	1	1	-1	-1
75	$-\nu$	$-\nu$	$-\nu$	1	1	1	0	-1

n=4

s = 3

Example 5.5

	1	3	5	8	9	15	35	59	84
1	0	-1	-1	-1	-1	ν	ν	ν	ν
2	1	-1	-1	-1	-1	-1	ν	ν	ν
5	1	1	0	-1	-1	-1	-1	ν	ν
8	1	1	1	0	-1	-1	-1	-1	ν
9	1	1	1	1	0	-1	-1	-1	-1
28	$-\nu$	1	1	1	1	1	-1	-1	-1
36	$-\nu$	$-\nu$	1	1	1	1	1	-1	-1
59	$-\nu$	$-\nu$	$-\nu$	1	1	1	1	0	-1
85	$-\nu$	$-\nu$	$-\nu$	$-\nu$	1	1	1	1	1

n=4

s = 4

In order to reduce the scope of our study somewhat, we shall restrict ourselves in the remainder of the paper to balanced games, defined as follows:

DEFINITION 5.6. Let $\tilde{W}_1$ and $\tilde{W}_2$ be pre-essential strategy sets. The game on $\tilde{W}_1 \times \tilde{W}_2$ is called <u>balanced</u> provided that n = s and there are no zeros off the diagonal in the payoff matrix.

Example 5.5 above is balanced, but 5.4 is not. The payoff matrix for a balanced game is completely determined by the diagonal, and the off-diagonal part is skew-symmetric. Since interchanging strategy sets changes the matrix to its negative transposed, we may assume without loss of generality that the first nonzero diagonal element is -1. Note also that invariance under step B implies that 1 and -1 do not occur consecutively on the diagonal, but must always be separated by a zero.

The case n = 0 is trivial. In the next section we discuss the case n = 1.

6. Balanced 3 by 3 games.

When $n = 1$ the pre-essential sets have three elements each. There are nine different possible diagonals, and none of these games reduces further. Thus $\tilde{W}_1$ and $\tilde{W}_2$ are already the essential sets. The nine diagonals and the solutions of the corresponding 3 by 3 games are given below. We abbreviate the diagonal elements -1 and +1 by - and +, respectively. $P = (p_1, p_2, p_3)$ is the optimal strategy for Player I, $Q = (q_1, q_2, q_3)$ that for Player II. V is the game value.

1. 000. This is the symmetric game, and the solution, as given in [2], is $P = Q = (1,\nu,1)/(\nu+2)$; $V = 0$.

2. 00-. $P = (\nu+3, \nu^2+2\nu-1, \nu+2)/(\nu+2)^2$, $Q = (\nu+1, (\nu+1)^2, \nu+2)/(\nu+2)^2$; $V = -1/(\nu+2)^2$.

3. 0-0. $P = (2, \nu^2+2\nu, 2\nu+2)/(\nu+2)^2$, $Q = (2\nu+2, \nu^2+2\nu, 2)/(\nu+2)^2$, $V = -\nu^2/(\nu+2)^2$.

4. 0--. $P = (4, \nu^2+2\nu-1, 2\nu+2)/(\nu^2+4\nu+5)$, $Q = (2\nu+2, (\nu+1)^2, 2)/(\nu^2+4\nu+5)$; $V = -(\nu^2+1)/(\nu^2+4\nu+5)$.

5. -0+. $P = (1, \nu+1, 1)/(\nu+3)$, $Q = (1, \nu-1, 1)/(\nu+1)$, $V = 0$.

6. -00. $P = (\nu+2, (\nu+1)^2, \nu+1)/(\nu+2)^2$, $Q = (\nu+2, \nu^2+2\nu-1, \nu+3)/(\nu+2)^2$; $V = -1/(\nu+2)^2$.

7. -0-. $P = (2,\nu^2+2\nu,2\nu+2)/(\nu+2)^2$, $Q = (2\nu+2,\nu^2+2\nu,2)/(\nu+2)^2$; $V = -\nu^2/(\nu+2)^2$.

8. --0. $P = (2,(\nu+1)^2,2\nu+2)/(\nu^2+4\nu+5)$, $Q = (2\nu+2,\nu^2+2\nu-1,4)/(\nu^2+4\nu+5)$; $V = -(\nu^2+1)/(\nu^2+4\nu+5)$.

9. ---. $P = (\alpha^2,1,\alpha)/(1+\alpha+\alpha^2)$, $Q = (\alpha,1,\alpha^2)/(1+\alpha+\alpha^2)$; $V = (-1+\alpha-\alpha^2)/(1+\alpha+\alpha^2)$, where $\alpha=2/(\nu+1)$. Here $\tilde{W}_1$ and $\tilde{W}_2$ are disjoint, and the reduced game is in the Class 4B.1 of [7].

There is a duality in cases (2) and (6) and again in the pair (4) and (8). In each pair, the diagonal of one is the reverse of that of the other. The vector P in one is the reverse of Q in the other, and the game values are equal. The reason is easy to see. The game matrix in (2) is $\begin{bmatrix} 0 & -1 & \nu \\ 1 & 0 & -1 \\ -\nu & 1 & -1 \end{bmatrix}$, so P must satisfy the inequalities

$$\begin{aligned} p_2 - \nu p_3 &\geq V \\ -p_1 + p_3 &\geq V \\ \nu p_1 - 1p_2 - p_3 &\geq V. \end{aligned}$$

The matrix in game (6) is $\begin{bmatrix} -1 & -1 & \nu \\ 1 & 0 & -1 \\ -\nu & 1 & 0 \end{bmatrix}$, so Q in this game must satisfy the inequalities

$$\begin{aligned} q_2 - \nu q_1 &\le V \\ -q_3 \qquad + \ q_1 &\le V \\ \nu q_3 - q_2 - q_1 &\le V. \end{aligned}$$

Since all three strategies are essential, i.e., no components may be zero, equality must hold throughout, and thus (q_3, q_2, q_1) must satisfy the same equations that (p_1, p_2, p_3) does.

7. Balanced 5 by 5 games.

Subject to our restriction that the first nonzero diagonal element is -, there are exactly 50 balanced 5 by 5 games. We may list them in lexicographic order of diagonals from 0 0 0 0 0 to - - - - - (with the ordering 0 < - < +). Of these fifty, the five with diagonals of the form - 0 + x y reduce to 2 by 2 games of type A, as may be seen from Theorem 10.1 below. They are numbers 34-38 in our ordering. The four with diagonals x y - 0 + similarly reduce to 2 by 2 games of type A', as implied by Theorem 10.2. They are numbers 7, 19, 31 and 48. The four having diagonals - x 0 y +, numbers 24, 28, 41 and 45, reduce to 3 by 3, as implied by Theorem 8.1.

In the remaining 37 games, it appears that all five pure strategies are essential; i.e., the essential sets are $W_1 = \tilde{W}_1$ and $W_2 = \tilde{W}_2$. The first, with diagonal 0 0 0 0 0, is the symmetric game; its solution is given in [2]. The last, with diagonal - - - - -, is the disjoint game of class 4B.2 in [7]. In Section 12 we give explicit solutions for a few further classes of games, of which some of the 5 by 5 games are special cases. As discussed in the last paragraph of Section 6, the games fall to some extent into pairs in which the solution for one member of the pair may be obtained immediately from that for the other.

There are several types of balanced 2n+1 by 2n+1 games that reduce to 5 by 5. These are special cases of balanced games that reduce to odd order, and we examine these in the next section.

8. Reduction of balanced games to odd order.

Recall that for balanced Silverman games the payoff matrix is completely determined by the diagonal, and that every diagonal element is 1, 0 or -1. The evidence strongly suggests that unless both 1 and -1 occur (and therefore all three of 1, 0, -1), the game is irreducible. If both 1 and -1 occur, with one of them in the middle position, then the game reduces to 2 by 2, as we show in Section 10. In this section and the next three, we examine the reduction for all other diagonals; i.e. those where each of 1, 0 and -1 occur on the diagonal and the middle element is 0. Those which reduce to an odd order game are treated in the present section and those reducing to even order in Section 9.

We shall refer to the first n diagonal elements as the left part and the last n elements as the right part, and we suppose now that these are separated by a central zero. Suppose at first that each of the left and right parts includes a nonzero element. Let a be the number of initial zeros in the left part and b be the number of final zeros in the left part. Similarly, let c and d be the numbers of initial and final zeros, respectively, in the right part. If we denote a string of u zeros by 0^u, the diagonals we are now considering have the form

(8.0.1) $\quad 0^a \quad w \quad G \quad x \quad 0^b \quad \boxed{0} \quad 0^c \quad y \quad H \quad z \quad 0^d$,

where each of w,x,y,z is 1 or -1, and G and H are arbitrary strings. The box indicates the middle element. We note that

(8.0.2) $a+b \le n-1$, with equality iff G is empty and w and x coincide;

$c+d \le n-1$, with equality iff H is empty and y and z coincide.

There are 16 possible sequences wxyz, but since interchanging roles of the two players changes the sign of each diagonal element, there is no loss of generality in assuming that w = -1, as we shall usually do. This leaves us with eight sequences, which we number as follows:

(8.0.3)

(i)	- - + +	(v)	- + + +
(ii)	- - + -	(vi)	- + + -
(iii)	- - - +	(vii)	- + - +
(iv)	- - - -	(viii)	- + - -

The notation (i') refers to the opposite sequence + + - -, and similarly for (ii'), etc. The games break further into cases as follows:

(8.0.4) (A) $a \le c,\ b \ge d$ (C) $a \le c,\ b < d$

(B) $a > c,\ b \ge d$ (D) $a > c,\ b < d$.

Sixteen of the resulting 32 cases reduced to balanced games (hence, odd order). The other sixteen reduce to even order games with some off-diagonal zeros.

Consider now diagonals in which one of the parts (left or right) consists entirely of zeros. We may represent these in the form

(8.0.5) $0^n\ \boxed{0}\ 0^c\ y\ H\ z\ 0^d$, or

(8.0.6) $0^a\ w\ G\ x\ 0^b\ \boxed{0}\ 0^n$.

Assuming again that the first nonzero diagonal element is -1, we have the cases

(8.0.7) (ix) 0 0 - - (xi) - - 0 0

(x) 0 0 - + (xii) - + 0 0 ,

with no further breakdown of the kind in (8.0.4). Two of these cases reduce to balanced (odd order) games, the other two to even order games with some off-diagonal zeros.

If $\nu > 1$ all of these reduced games appear not to be further reducible. But if $\nu = 1$ there is always a further reduction to a 2 by 2 game with matrix $\begin{bmatrix} -1 & 1 \\ 1 & -1 \end{bmatrix}$ or its negative.

The eighteen cases which reduce to odd order are (iA), (iB), (iC), (iD), (iiC), (iiD), (iiiA), (iiiC), (ivB), (ivC), (vA), (vB), (viiA), (viiD), (viiiB), (viiiD), (ix) and (xi). The reduced game is in each case a balanced game with one of the following diagonal types, or one of these with the roles of the players reversed:

(8.0.5A) $\quad 0^a \quad - \quad 0^d \quad \boxed{0} \quad 0^a \quad + \quad 0^d$

(8.0.5B) $\quad 0^{c+1} \quad - \quad 0^d \quad \boxed{0} \quad 0^c \quad + \quad + \quad 0^d$

(8.0.5C) $\quad 0^a \quad - \quad - \quad 0^b \quad \boxed{0} \quad 0^a \quad + \quad 0^{b+1}$

(8.0.5D) $\quad 0^{c+1} \quad - \quad 0^b \quad \boxed{0} \quad 0^c \quad + \quad 0^{b+1}$

The A, B, C and D in these labels correspond to the subclasses in (8.0.4). Thus, cases (iA), (iiiA), (vA) and (viiA) all reduce to type (8.0.5A), etc.

Our first theorem of this section deals with (iA), (iB), (vA), (vB), (iiiA) and (viiA). Let $t = \min\{a, c+1\}$,

$$W_1^1 = \{e_i: \quad 1 \le i \le t+1\},$$

$$W_1^2 = \{e_i: \quad n+1-d \le i \le n+t+1\},$$

$$W_1^3 = \{e_i: \quad 2n+1-d \le i \le 2n+1\},$$

$$W_2^1 = \{f_j: \quad 1 \le j \le t\} \cup \{f_{a+1}\},$$

$$W_2^2 = \{f_j: \quad n+1-d \le j \le n+t+1\} \cup \{f_{n+a+2}\},$$

$$W_2^3 = \{f_j: \quad 2n+2-d \le j \le 2n+1\}.$$

THEOREM 8.1 Assume that $b \geq d$, $w = -1$, $z = 1$, and, in case $a > c$, that $y = 1$. Let $W_1 = W_1^1 \cup W_1^2 \cup W_1^3$ and $W_2 = W_2^1 \cup W_2^2 \cup W_2^3$. Then optimal strategies for the (2t+2d+3) by (2t+2d+3) game on $W_1 \times W_2$ are optimal for the full game on $\hat{W}_1 \times \hat{W}_2$. The reduced game is the balanced game with diagonal (8.0.5A) if $a \leq c$, and (8.0.5B) if $a > c$.

PROOF. It will be helpful in reading the proof to refer to the payoff matrix in Figure 1. We show first that against W_2, each e_i in $\hat{W}_1 \setminus W_1$ is dominated by one in W_1, as follows:

(i) e_{t+1} dominates e_i for $t+1 \leq i \leq a+1$;

(ii) e_{n+1-d} dominates e_i for $a+2 \leq i \leq n+1-d$;

(iii) e_{n+t+1} dominates e_i for $n+t+1 \leq i \leq n+a+1$;

(iv) e_{2n+1-d} dominates e_i for $n+a+2 \leq i \leq 2n+1-d$.

For (i), let $t+1 \leq i \leq a+1$, and consider first such e_i against f_j in W_2^1. For $1 \leq j \leq t$ we have $j < i \leq a+1 \leq n < j+n$, so $a_{i,j} = 1$ in every case. Against f_{a+1} these e_i are likewise equivalent, since $a_{i,a+1} = -1$ when $t+1 \leq i < a+1$, and $a_{a+1,a+1} = -1$ by hypothesis. For such e_i against f_j in W_2^2, consider first $n+1-d \leq j \leq n+t+1$. From (8.0.1) we have $i < j \leq n+t+1 \leq i+n$, so each $a_{i,j} = -1$. Since $n+a+2 >$

		*	…	*			*			*	…	*			*			…	*
		f_1	…	f_t	f_{t+1}	…	f_{a+1}	…	f_{n-d}	f_{n+1-d}	…	f_{n+t+1}	f_{n+t+2}	…	f_{n+a+2}	…	f_{2n+1-d}	…	f_{2n+1}
*	e_1	0	…	−1	−1	…	−1	…	−1	−1	…	ν	ν	…	ν	…	ν	…	ν
⋮	⋮	⋮																	
*	e_t	1	…	0	−1	…	−1	…	−1	−1	…	ν	ν	…	ν	…	ν	…	ν
*	e_{t+1}	1	…	1	h	…	−1	…	−1	−1	…	−1	ν	…	ν	…	ν	…	ν
	⋮	⋮																	
	e_{a+1}	1	…	1	1	…	−1	…	−1	−1	…	−1	−1	…	ν	…	ν	…	ν
	⋮	⋮																	
	e_{n-d}	1	…	1	1	…	1	…	k	−1	…	−1	−1	…	−1	…	ν	…	ν
*	e_{n+1-d}	1	…	1	1	…	1	…	1	0	…	−1	−1	…	−1	…	−1	…	ν
⋮	⋮	⋮																	
*	e_{n+t+1}	$-\nu$	…	$-\nu$	1	…	1	…	1	1	…	m	−1	…	−1	…	−1	…	−1
	e_{n+t+2}	$-\nu$	…	$-\nu$	$-\nu$	…	1	…	1	1	…	1	n	…	−1	…	−1	…	−1
	⋮	⋮																	
	e_{n+a+2}	$-\nu$	…	$-\nu$	$-\nu$	…	$-\nu$	…	1	1	…	1	1	…	p	…	−1	…	−1
	⋮	⋮																	
*	e_{2n+1-d}	$-\nu$	…	$-\nu$	$-\nu$	…	$-\nu$	…	$-\nu$	1	…	1	1	…	1	…	1	…	−1
⋮	⋮	⋮																	
*	e_{2n+1}	$-\nu$	…	$-\nu$	$-\nu$	…	$-\nu$	…	$-\nu$	$-\nu$ or 1	…	1	1	…	1	…	1	…	q

Figure 1. Game matrix for Theorem 8.1.
$h,k,m,n,p,q \in \{-1,0,1\}$

$i+n$, each $e_{i,n+a+2} = \nu$, and thus all e_i in this group are equivalent against W_2^2. If f_j is in W_2^3 we have $j \geq 2n+2-d$, while $i \leq a+1 < n+1-b \leq n+1-d$, so $j > i+n$ and $a_{i,j} = \nu$ in every case. Thus all e_i in this group are equivalent against W_2.

(ii) Let $a+2 \leq i \leq n+1-d$. For f_j in W_2^1 we have $j \leq a+1 < i \leq n+1 \leq n+j$, so every $a_{i,j} = 1$. For f_j in W_2^2 and such i we have $i \leq j \leq n+a+2 \leq i+n$. If $i < j$ then $a_{i,j} = -1$, and if $i = j = n+1-d$ then $a_{i,j} = 0$ since $d \leq b$. Thus e_{n+1-d} dominates.

(iii) Let $n+t+1 \leq i \leq n+a+1$. If $t = a$ then e_{n+t+1} is the only e_i in this range, and there is nothing to prove, so assume that $t = c+1 < a$. For f_j in $W_2^1 \setminus \{f_{a+1}\}$ we have $i > j+n$, so that every $a_{i,j} = -\nu$. For f_j in $\{f_{a+1}\} \cup W_2^2 \setminus \{f_{n+a+2}\}$ we have $j \leq i \leq n+a+1 \leq j+n$. If $j < i$ then $a_{i,j} = 1$. If $i = j = n+t+1 = n+c+2$, then $a_{i,j} = y = 1$ by hypotheses. For $n+a+2 \leq j \leq 2n+1$ we have $i < j \leq i+n$, and hence every $a_{i,j} = -1$. Thus all e_i in this group are equivalent against W_2.

(iv) Let $n+a+2 \leq i \leq 2n+1-d$. For all $j \leq a+1$ we have $i > j+n$ and thus $a_{i,j} = -\nu$. For $n+1-d \leq j \leq n+t+1$ we have $j < i \leq j+n$, so that $a_{i,j} = 1$. If $j = n+a+2$ then $j \leq i < j+n$. When $j = i$, $a_{i,j} \leq 1$; in all other cases $a_{i,j} = 1$. In particular $a_{2n+1-d,j} = 1 \geq a_{i,j}$ for all i in this range. For $2n+2-d \leq j \leq 2n+1$ we have $i < j < i+n$ and hence $a_{i,j} = -1$. Thus against W_2, e_{2n+1-d} dominates all e_i in this group.

To complete the proof we show that against W_1 each f_j in $\tilde{W}_2 \setminus W_2$ is dominated by one in W_2, as follows.

(i) f_{a+1} dominates f_j for $t+1 \leq j \leq n-d$.

(ii) f_{n+a+2} dominates f_j for $n+t+2 \leq j \leq 2n+1-d$.

For (i), let $t+1 \leq j \leq n-d$, and consider first such f_j against W_1^1. Then $i \leq t+1 \leq j \leq n-d < n+i$. If $i < j$ then $a_{i,j} = -1$. If $i = j = t+1$ and $t < a$ then $a_{i,j} = 0$ while $a_{i,a+1} = -1$. If $i = j = t+1 = a+1$ then $a_{i,j} = -1$ by hypothesis. Thus, against W_1^1, f_{a+1} dominates the f_j in this group. For e_i in W_1^2 and such j we have $j < i \leq j+n$, so every $a_{i,j} = 1$. For e_i in W_1^3 and such j we have $i > j+n$, and every $a_{i,j} = -\nu$. Thus f_{a+1} dominates against all of W_1.

(ii) Let $n+t+2 \leq j \leq 2n+1-d$. For e_i in W_1^1 we have $j > i+n$, so $a_{i,j} = \nu$. For e_i in W_1^2, $i < j \leq i+n$, whence $a_{i,j} = -1$ in every case. For e_i in W_1^3, $j \leq i < j+n$. If $j < i$ then $a_{i,j} = 1$ in every case, and if $j = i = 2n+1-d$ then $a_{i,j} = 1$ by hypothesis. Thus all f_j in this range are in fact equivalent against W_1, and the proof is complete. (It is easy to check that the reduced game has the diagonal asserted.) □

The next theorem deals with cases (iC), (iD), (iiC), (iiD), (viiD) and (viiiD). Let u = min {a+1,c+1}, and define the sets

$$W_1^1 = \{e_i: \quad 1 \le i \le u\} \cup \{e_{c+2}\},$$

$$W_1^2 = \{e_i: \quad n+1-b \le i \le n+u\} \cup \{e_{n+c+2}\},$$

$$W_1^3 = \{e_i: \quad 2n+1-b \le i \le 2n+1\},$$

$$W_2^1 = \{f_j: \quad 1 \le j \le u\},$$

$$W_2^2 = \{f_j: \quad n-b \le j \le n+1+u\},$$

$$W_2^3 = \{f_j: \quad 2n+1-b \le j \le 2n+1\}.$$

Cases (viiD) and (viiiD) are settled in this theorem by observing that when $a > c$ and $b < d$, the proof is valid also when w (the diagonal element following the initial a zeros) is +1. This means that the reduction is valid for (vii') + - + - and (viii') + - + + in case (D), so that by interchanging W_1 and W_2 we have reduced optimal sets for (vii) - + - + and (viii) - + - -.

THEOREM 8.2. Assume that $b < d$, $x = -1$ and $y = +1$. We assume $w = -1$ only in case $a \le c$. With W_i^j as defined in the preceding paragraph, let $W_i = W_i^1 \cup W_i^2 \cup W_i^3$, $i = 1,2$. Then optimal strategies for the

(2u+2b+3) by (2u+2b+3) game on $W_1 \times W_2$ are optimal for the full game on $\tilde{W}_1 \times \tilde{W}_2$. The reduced game is the balanced game with diagonal (8.0.5C) if $a \le c$. In cases (iD) and (iiD) the reduced game is the balanced game with diagonal (8.0.5D) and in (viiD) and (viiiD) it is that with diagonal (8.0.5D'), namely $0^{c+1} + 0^b \boxed{0}\ 0^c - 0^{b+1}$.

PROOF. The game matrix is shown in Figure 2. We show first that against W_2, each e_i in $\tilde{W}_1 \setminus W_1$ is dominated by one in W_1, as follows:

(i) e_{c+2} dominates e_i for $u+1 \le i \le n-b$;

(ii) e_{n+c+2} dominates e_i for $n+u+1 \le i \le 2n-b$.

For (i), let $u+1 \le i \le n-b$, and consider first such e_i against f_j in W_2^1. Since $1 \le j \le u$ we have $j < i < j+n$, so each $a_{i,j} = 1$. Next, if $f_j \in W_2^2$ we have $n-b \le j \le n+1+u$, so that $i \le j \le i+n$. If $i < j$, each $a_{i,j} = -1$, and if $i = j = n-b$ then $a_{i,j} = x = -1$ by hypothesis. Consider f_j in W_2^3. Then $j \ge 2n+1-b > i+n$, so each $a_{i,j} = \nu$. Thus all e_i in this group are equivalent against W_2.

(ii) Let $n+u+1 \le i \le 2n-b$. For $1 \le j \le u$ we have $i > j+n$, so every $a_{i,j} = -\nu$. For $n-b \le j \le n+1+u$ we have $j \le i \le j+n$. If $j < i$, every $a_{i,j} = 1$. If $j =$

		f_1	…	f_u	…	f_{c+2}	…	f_{n-b}	f_{n+1-b}	…	f_{n+u}	f_{n+u+1}	…	f_{n+c+2}	…	f_{2n-b}	f_{2n+1-b}	…	f_{2n+1}
		*	…	*				*	*	…	*	*					*	…	*
*	e_1	0	…	−1	…	−1	…	−1	−1	…	ν	ν	…	ν	…	ν	ν	…	ν
*	e_u	1	…	h	…	−1	…	−1	−1	…	−1	ν	…	ν	…	ν	ν	…	ν
*	e_{c+2}	1	…	1	…	k	…	−1	−1	…	−1	−1	…	−1	…	ν	ν	…	ν
	e_{n-b}	1	…	1	…	1	…	−1	−1	…	−1	−1	…	−1	…	−1	ν	…	ν
*	e_{n+1-b}	1	…	1	…	1	…	1	0	…	−1	−1	…	−1	…	−1	−1	…	ν
*	e_{n+u}	$-\nu$	…	1	…	1	…	1	1	…	m	−1	…	−1	…	−1	−1	…	−1
	e_{n+u+1}	$-\nu$	…	$-\nu$	…	1	…	1	1	…	1	n	…	−1	…	−1	−1	…	−1
*	e_{n+c+2}	$-\nu$	…	$-\nu$	…	1	…	1	1	…	1	1	…	1	…	−1	−1	…	−1
	e_{2n-b}	$-\nu$	…	$-\nu$	…	$-\nu$	…	1	1	…	1	1	…	1	…	p	−1	…	−1
*	e_{2n+1-b}	$-\nu$	…	$-\nu$	…	$-\nu$	…	$-\nu$	1	…	1	1	…	1	…	1	0	…	−1
*	e_{2n+1}	$-\nu$	…	$-\nu$	…	$-\nu$	…	$-\nu$	$-\nu$	…	1	1	…	1	…	1	1	…	0

Figure 2. Game matrix for Theorem 8.2
Diagonal elements h,k,m,n,p $\in$ {-1,0,1}

$i = n+u+1$, then $a_{i,j} \le 1 = a_{n+c+2,j}$, so against $W_2^1 \cup W_2^2$ e_{n+c+2} dominates all e_i in this group. For f_j in W_2^3 we have $2n+1-b \le j \le 2n+1$, so $i < j < i+n$, and each $a_{i,j} = -1$. Thus e_{n+c+2} dominates in this group against every f_j in W_2.

To complete the proof we show that against W_1, each f_j in $\widetilde{W}_2 \setminus W_2$ is dominated by one in W_2, as follows.

(i) f_u dominates f_j for $u \le j \le c+1$;

(ii) f_{n-b} dominates f_j for $c+2 \le j \le n-b$;

(iii) f_{n+u+1} dominates f_j for $n+u+1 \le j \le n+c+2$;

(iv) f_{2n+1-b} dominates f_j for $n+c+3 \le j \le 2n+1-b$.

For (i), let $u \le j \le c+1$. If $a \ge c$ then $u = c+1$ and there is nothing to prove. Thus, suppose $a < c$, and consider first e_i with $i \le u$, so that $i \le j \le i+n$. If $i < j$ we have $a_{i,j} = -1$, and if $i = j = u$, then since $u = a+1$ we have $a_{i,j} = w = -1$ by hypothesis, so against these e_i, all f_j in this group are equivalent. Next consider e_i with $c+2 \le i \le n+u$. Then $j < i \le j+n$, so each $a_{i,j} = 1$. For all $i \ge n+c+2$ we have $i > j+n$ and hence $a_{i,j} = -\nu$. Thus all f_j in this group are equivalent against all e_i in W_1.

(ii) Let $c+2 \le j \le n-b$, and consider first e_i in W_1^1. Thus $1 \le i \le c+2$. In view of (8.0.1) we have

$c+2 \le n-d+1 \le n-b$. For $i < c+2$ then, $a_{i,n-b} = -1$. If $c+2 < n-b$, then $a_{c+2,n-b} = -1$ also, and $a_{n-b,n-b} = x = -1$ by hypothesis, so we have $a_{i,n-b} = -1 \le a_{i,j}$ for all i,j under consideration. Next consider e_i in W_1^2. Then $n+1-b \le i \le n+c+2 \le 2n-b$, so $j < i \le j+n$, and each $a_{i,j} = 1$. Now consider e_i in W_1^3. Then $i > j+n$, so each $a_{i,j} = -\nu$. Thus, against all e_i in W_1, f_{n-b} dominates the f_j in this group.

(iii) Let $n+1+u \le j \le n+c+2$. If $u = c+1$ there is nothing to prove here, so we may assume $u = a+1 < c+1$. For $1 \le i \le u$ we have $j > i+n$, and every $a_{i,j} = \nu$. For $c+2 \le i \le n+u$ we have $i < j \le i+n$, so each $a_{i,j} = -1$. For $n+c+2 \le i \le 2n+1$, $j \le i < j+n$. If $j < i$ then $a_{i,j} = 1$, and if $j = i = n+c+2$, then $a_{i,j} = y = 1$ by hypothesis. Thus, against W_1, all f_j in this group are equivalent.

(iv) Let $n+c+3 \le j \le 2n+1-b$. For $1 \le i \le c+2$, every $a_{i,j}$ is ν, since $j > i+n$. For $n+1-b \le i \le n+c+2$ we have $i < j \le i+n$, so each $a_{i,j} = -1$. For $2n+1-b \le i \le 2n+1$ we have $j \le i < j+n$. If $j < i$ then $a_{i,j} = 1$. If $j = i = 2n+1-b$ then $a_{i,j} = 0$ because $b < d$. Thus $a_{i,2n+1-b} \le a_{i,j}$ for all j in this group and all e_i in W_1, so the proof is complete. □

The next theorem takes care of the single case (viiiB), - + - - with $a > c$, $b \geq d$. For this theorem we define

$$W_1^1 = \{e_i: \quad 1 \leq i \leq c+1\},$$

$$W_1^2 = \{e_{n-b}\} \cup \{e_i: \quad n+1-d \leq i \leq n+c+2\},$$

$$W_1^3 = \{e_{2n+1-b}\} \cup \{e_i: \quad 2n+2-d \leq i \leq 2n+1\},$$

$$W_2^1 = \{f_j: \quad 1 \leq j \leq c+2\},$$

$$W_2^2 = \{f_j: \quad n+1-d \leq j \leq n+c+2\},$$

$$W_2^3 = \{f_j: \quad 2n+1-d \leq j \leq 2n+1\}.$$

THEOREM 8.3. Assume that $a > c$, $b \geq d$, $x = 1$ and $y = z = -1$. With W_i^j as defined above, let $W_i = W_i^1 \cup W_i^2 \cup W_i^3$, $i = 1,2$. Then optimal strategies for the $(2c+2d+5)$ by $(2c+2d+5)$ game on $W_1 \times W_2$ are optimal for the full game on $\widetilde{W}_1 \times \widetilde{W}_2$. The reduced game is the balanced game with diagonal (8.0.5B').

PROOF. The game matrix is shown in Figure 3. We show first that against W_2, each e_i in $\widetilde{W}_1 \setminus W_1$ is dominated by one in W_1, as follows:

(i) e_{n-b} dominates e_i for $c+2 \leq i \leq n-d$;

(ii) e_{2n+1-b} dominates e_i for $n+c+3 \leq i \leq 2n+1-d$.

For (i), let $c+2 \leq i \leq n-d$, and consider first such e_i against f_j in W_2^1. Then $j \leq i \leq n+j$, so every $a_{i,j} \leq 1$, with $a_{i,j} = 1$ when $i > j$ and $a_{c+2,c+2} = 1$, 0

		*	...	*	*	...		...	*	...	*	...		...	*	...	*
		f_1	...	f_{c+1}	f_{c+2}	...	f_{n-b}	...	f_{n+1-d}	...	f_{n+c+2}	...	f_{2n+1-b}	...	f_{2n+1-d}	...	f_{2n+1}
*	e_1	0	...	−1	−1	...	−1	...	−1	...	ν	...	ν	...	ν	...	ν
⋮	⋮	⋮															
*	e_{c+1}	1	...	0	−1	...	−1	...	−1	...	ν	...	ν	...	ν	...	ν
	e_{c+2}	1	...	1	h	...	−1	...	−1	...	−1	...	ν	...	ν	...	ν
	⋮	⋮															
*	e_{n-b}	1	...	1	1	...	1	...	−1	...	−1	...	ν	...	ν	...	ν
	⋮	⋮															
*	e_{n+1-d}	1	...	1	1	...	1	...	0	...	−1	...	−1	...	−1	...	ν
⋮	⋮	⋮															
*	e_{n+c+2}	$-\nu$	...	$-\nu$	1	...	1	...	1	...	−1	...	−1	...	−1	...	−1
	⋮	⋮															
*	e_{2n+1-b}	$-\nu$	...	$-\nu$	$-\nu$	...	$-\nu$	...	1	...	1	...	k	...	−1	...	−1
	⋮	⋮															
	e_{2n+1-d}	$-\nu$	...	$-\nu$	$-\nu$	...	$-\nu$	...	1	...	1	...	1	...	−1	...	−1
	⋮	⋮															
*	e_{2n+1}	$-\nu$	...	$-\nu$	$-\nu$	...	$-\nu$	...	m	...	1	...	1	...	1	...	p

Figure 3. Game matrix for Theorem 8.3.
Diagonal elements h,k,p are 0 or ±1.

or -1. Note that with $a > c$, (8.0.1) implies $n-b \geq c+2$. If $n-b > c+2$ then $a_{n-b,j}$ is still 1 for every j since $a_{n-b,n-b} = x = 1$ by hypothesis. Thus, against W_2^1, e_{n-b} dominates the e_i in this group. For f_j in W_2^2 we have $i < j \leq i+n$, so every $a_{i,j} = -1$. For f_j in W_2^3, $j > i+n$ and every $a_{i,j} = \nu$. Thus against $W_2^2 \cup W_2^3$ all e_i in this group are equivalent.

(ii) Let $n+c+3 \leq i \leq 2n+1-d$, and consider first such e_i against f_j in W_2^1. Since $i > j+n$, every $a_{i,j} = -\nu$. For f_j in W_2^2, $j < i \leq j+n$, so every $a_{i,j} = 1$. For f_j in W_2^3 we have $i \leq j < i+n$. If $i < j$ then $a_{i,j} = -1$. If $i = j = 2n+1-d$ then $a_{i,j} = z = -1$. Thus all e_i in this group are equivalent against W_2.

We complete the proof by showing that against W_1, each f_j in $\widetilde{W}_2 \setminus W_2$ is dominated by one in W_2, as follows:

(i) f_{c+2} dominates f_j for $c+2 \leq j \leq n-b$;

(ii) f_{n+1-d} dominates f_j for $n+1-b \leq j \leq n+1-d$;

(iii) f_{n+c+2} dominates f_j for $n+c+2 \leq j \leq 2n-b$;

(iv) f_{2n+1-d} dominates f_j for $2n+1-b \leq j \leq 2n+1-d$.

For (i), let $c+2 \leq j \leq n-b$, and consider such f_j against e_i in W_1^1. Then $1 < j < i+n$ so that every $a_{i,j} = -1$. For e_i in W_1^2 we have $j \leq i \leq j+n$. If $j < i$

then $a_{i,j} = 1$, and if $j = i = n-b$ then $a_{i,j} = x = 1$. For e_i in W_1^3, $i > j+n$, and every $a_{i,j} = -\nu$. Thus all f_j in this range are equivalent against W_1.

(ii) Let $n+1-b \le j \le n+1-d$, and consider first such f_j against e_i with $1 \le i \le n-b$. Then $1 < j \le i+n$, so every $a_{i,j} = -1$. Next consider such f_j against e_i with $n+1-d \le i \le 2n+1-b$. Then $j \le i \le j+n$. For $j < i$, each $a_{i,j} = 1$, and if $j = i = n+1-d$ then $a_{i,j} = 0$ because $b \ge d$. Thus f_{n+1-d} dominates against e_i in this range. For e_i with $2n+2-d \le i \le 2n+1$ we have $i > j+n$, so every $a_{i,j} = -\nu$. Thus against all e_i in W_1, f_{n+1-d} dominates the f_j in this group.

(iii) Let $n+c+2 \le j \le 2n-b$, and consider first such f_j against e_i in W_1^1. Then $j > i+n$, so every $a_{i,j} = \nu$. For e_i in W_1^2 we have $i \le j \le i+n$. If $i < j$ then $a_{i,j} = -1$, and if $i = j = n+c+2$ then $a_{i,j} = y = -1$ by hypothesis. For e_i in W_1^3, we have $i > j+n$, and every $a_{i,j} = -\nu$. Thus, against the e_i in W_1, all f_j in this group are equivalent.

(iv) Let $2n+1-b \le j \le 2n+1-d$, and consider first such f_j against e_i with $i \le n-b$. Then $j > i+n$, so every $a_{i,j} = \nu$. Next consider such f_j against e_i with $n+1-d \le i \le 2n+1-b$. Then $i \le j \le i+n$, and for $i < j$

each $a_{i,j} = -1$. If $i = j = 2n+1-b$, $a_{i,j} \geq -1$. Since $z = -1$, $a_{i,2n+1-d} = -1$ for all i in this range, and thus f_{2n+1-d} dominates. Finally, consider such f_j against e_i with $2n+2-d \leq i \leq 2n+1$. Then $j < i \leq j+n$, so every $a_{i,j} = 1$. Thus, against all e_i in W_1, f_{2n+1-d} dominates the f_j in this group, and the proof is complete. □.

The next theorem likewise treats a single case, namely (iiiC): $- - - +$ with $a \leq c$ and $b < d$. For this theorem we define the sets

$$W_1^1 = \{e_i: \quad 1 \leq i \leq a+1\},$$

$$W_1^2 = \{e_{n+1-d}\} \cup \{e_i: \quad n+1-b \leq i \leq n+a+1\},$$

$$W_1^3 = \{e_{2n+1-d}\} \cup \{e_i: \quad 2n+1-b \leq i \leq 2n+1\},$$

$$W_2^1 = \{f_j: \quad 1 \leq j \leq a+1\},$$

$$W_2^2 = \{f_j: \quad n-b \leq j \leq n+a+2\},$$

$$W_2^3 = \{f_j: \quad 2n+1-b \leq j \leq 2n+1\}.$$

THEOREM 8.4. Assume that $x = y = -1$, $z = 1$, $a \leq c$ and $b < d$. Let W_i^j be as defined above, and $W_i = W_i^1 \cup W_i^2 \cup W_i^3$, $i = 1,2$. Then optimal strategies for the (2a+2b+5) by (2a+2b+5) game on $W_1 \times W_2$ are optimal for the full game on $\widetilde{W}_1 \times \widetilde{W}_2$. The reduced game is the balanced game with diagonal (8.0.5C).

PROOF. The game matrix is shown in Figure 4. We show first that against W_2, each e_i in $\widetilde{W}_1 \setminus W_1$ is dominated by one in W_1, as follows:

		*	...	*				*	*	...	*	*				*	...	*
		f_1	...	f_{a+1}	...	f_{n+1-d}	...	f_{n-b}	f_{n+1-b}	...	f_{n+a+1}	f_{n+a+2}	...	f_{2n+1-d}	...	f_{2n+1-b}	...	f_{2n+1}
*	e_1	0	...	-1	...	-1	...	-1	-1	...	ν	ν	...	ν	...	ν	...	ν
⋮	⋮	⋮																
*	e_{a+1}	i	...	-1	...	-1	...	-1	-1	...	-1	ν	...	ν	...	ν	...	ν
	⋮	⋮																
*	e_{n+1-d}	i	...	1	...	h	...	-1	-1	...	-1	-1	...	-1	...	ν	...	ν
	⋮	⋮																
	e_{n-b}	i	...	1	...	1	...	-1	-1	...	-1	-1	...	-1	...	ν	...	ν
*	e_{n+1-b}	1	...	1	...	1	...	1	0	...	-1	-1	...	-1	...	-1	...	ν
⋮	⋮	⋮																
*	e_{n+a+1}	$-\nu$	...	1	...	1	...	1	1	...	0	-1	...	-1	...	-1	...	-1
	e_{n+a+2}	$-\nu$	...	$-\nu$	...	1	...	1	1	...	1	k	...	-1	...	-1	...	-1
	⋮	⋮																
*	e_{2n+1-d}	$-\nu$	...	$-\nu$	...	1	...	1	1	...	1	1	...	1	...	-1	...	-1
	⋮	⋮																
*	e_{2n+1-b}	$-\nu$	...	$-\nu$	...	$-\nu$	...	$-\nu$	1	...	1	1	...	1	...	0	...	-1
⋮	⋮	⋮																
*	e_{2n+1}	$-\nu$	...	$-\nu$	...	$-\nu$	...	$-\nu$	$-\nu$	...	1	1	...	1	...	1	...	0

(i) e_{n+1-d} dominates e_i for $a+2 \le i \le n-b$;

(ii) e_{2n+1-d} dominates e_i for $n+a+2 \le i \le 2n-b$.

For (i), let $a+2 \le i \le n-b$, and consider first such e_i against f_j in W_2^1. Then $j < i \le j+n$, so every $a_{i,j} = 1$. Next consider such e_i against f_j in W_2^2, where we have $i \le j \le i+n$. For $i < j$, each $a_{i,j} = -1$, and if $i = j = n-b$ then $a_{i,j} = x = -1$ by hypothesis. Finally, consider such e_i against f_j in W_2^3. Then $j > i+n$, so every $a_{i,j} = \nu$. Thus, against W_2, all e_i in this group are equivalent.

(ii) Let $n+a+2 \le i \le 2n-b$, and consider first such e_i against f_j in W_2^1. Since $i > j+n$, every $a_{i,j} = -\nu$. Next consider such e_i against f_j in W_2^2. Then $j \le i \le j+n$, so every $a_{i,j} \le 1$, with $a_{i,j} = 1$ when $i > j$. If $2n+1-d > n+a+2$ then every $a_{2n+1-d,j} = 1 \ge a_{i,j}$. If $i = 2n+1-d = n+a+2$ then $a_{i,i} = z = 1$ by hypothesis, so against W_2^2, e_{2n+1-d} dominates the e_i in this group. Lastly, consider such e_i against f_j in W_2^3. Then $i < j \le i+n$, so every $a_{i,j} = -1$. Thus, against all of W_2, e_{2n+1-d} dominates the e_i in this group.

We complete the proof by showing that against W_1, each f_j in $\hat{W}_2 \setminus W_2$ is dominated by one in W_2, as follows:

(i) f_{a+1} dominates f_j for $a+1 \le j \le n-d$;

(ii) f_{n-b} dominates f_j for $n+1-d \le j \le n-b$;

(iii) f_{n+a+2} dominates f_j for $n+a+2 \le j \le 2n+1-d$;

(iv) f_{2n+1-b} dominates f_j for $2n+2-d \le j \le 2n+1-b$.

For (i), let $a+1 \le j \le n-d$, and consider first such f_j against e_i in W_1^1. Then $i \le j \le i+n$. For $i < j$ every $a_{i,j} = -1$. If $i = j = a+1$ then $a_{i,j} = -1$ by hypothesis. Thus all f_j in this group are equivalent against W_1^1. Next consider such f_j against e_i in W_1^2. Then $j < i \le n+j$, so every $a_{i,j} = 1$. Finally, consider such f_j against e_i in W_1^3. Then $i > j+n$, so every $a_{i,j} = -\nu$. Thus the f_j in this group are equivalent against all e_i in W_1.

(ii) Let $n+1-d \le j \le n-b$, and consider first such f_j against e_i with $i \le n+1-d$. Note that from $a \le c$ and (8.0.1) we have $a+d \le c+d \le n-1$, so that $a+1 < n+1-d$. Since $i \le j \le i+n$, each $a_{i,j} \ge -1$. With $j = n-b$, each $a_{i,j} = -1$ (including $i = j$, since $x = -1$ by hypothesis), so f_{n-b} dominates. Next consider such f_j against e_i with $n+1-b \le i \le 2n+1-d$. Now $j < i \le j+n$, so every $a_{i,j} = 1$. Lastly, consider such f_j against e_i with $2n+1-b \le i \le 2n+1$. Then $i > j+n$, so every $a_{i,j} = -\nu$. Thus, against all e_i in W_1, f_{n-b} dominates in this group.

(iii) Let $n+a+2 \le j \le 2n+1-d$, and consider first such f_j against e_i in W_1^1. Then $j > i+n$, so every $a_{i,j} = \nu$. Next consider such f_j against e_i in W_1^2. Then $i < j \le i+n$, so every $a_{i,j} = -1$. Now consider such f_j against e_i in W_1^3. Then $j < i \le j+n$ and every $a_{i,j} = 1$. Thus all f_j in this group are equivalent against W_1.

(iv) Let $2n+2-d \le j \le 2n+1-b$, and consider first such f_j against e_i with $i \le n+1-d$. Then $j > i+n$, so every $a_{i,j} = \nu$. Next consider such f_j against e_i with $n+1-b \le i \le n+a+1$. As we saw in (ii), $a+1 < n+1-d$, so $i < j \le i+n$, and every $a_{i,j} = -1$. Finally, consider such f_j against e_i with $2n+1-b \le i$. Then $j \le i \le j+n$. If $j < i$, $a_{i,j} = 1$. If $j = i = 2n+1-b$ then, since $b < d$, we have $a_{i,j} = 0$. Thus, against these e_i, and hence against all e_i in W_1, f_{2n+1-b} dominates in this group. This completes the proof. □

We turn now to cases (iv), (ix) and (xi), where, as mentioned earlier, there appears to be no reduction unless +1 occurs somewhere in the string G or H in (8.0.1), (8.0.5) or (8.0.6). The cases where +1 is in G and where +1 is in H are treated separately. The following theorem deals with the first subcase, (ivBG). Note that since - and + on the diagonal must be separated by a 0, such a + can occur only in a position k for which $a+3 \le k \le n-b-2$.

THEOREM 8.5. Assume that $a > c$, $b \ge d$, $w = x = y = z = -1$, and that for some k with $a+3 \le k \le n-b-2$, +1 occurs on the diagonal in position k. Let

$W_1^1 = \{e_i: 1 \le i \le c+1\} \cup \{e_k\}$,

$W_1^2 = \{e_i: n+1-d \le i \le n+c+2\} \cup \{e_{n+k+1}\}$,

$W_1^3 = \{e_i: 2n+2-d \le i \le 2n+1\}$,

$W_2^1 = \{f_j: 1 \le j \le c+2\}$,

$W_2^2 = \{f_j: n+1-d \le j \le n+c+2\}$,

$W_2^3 = \{f_j: 2n+1-d \le j \le 2n+1\}$,

and $W_i = W_i^1 \cup W_i^2 \cup W_i^3$ for $i = 1,2$. Then mixed strategies which are optimal for the (2c+2d+5) by (2c+2d+5) subgame on $W_1 \times W_2$ are optimal for the full game on $\widetilde{W}_1 \times \widetilde{W}_2$. The reduced game is the balanced game with diagonal (8.0.5B').

PROOF. The proof is indicated by the game matrix in Figure 5. We show first that against W_2, every e_i in $\widetilde{W}_1 \setminus W_1$ is dominated by one in W_1, as follows:

(i) e_k dominates e_i for $c+2 \le i \le n-d$, and

(ii) e_{n+k+1} dominates e_i for $n+c+3 \le i \le 2n+1-d$.

For (i), let $c+2 \le i \le n-d$, and consider first

		*	...	*	*				*	...	*				*	*	...	*
		f_1	...	f_{c+1}	f_{c+2}	...	f_k	...	f_{n+1-d}	...	f_{n+c+2}	...	f_{n+k+1}	...	f_{2n+1-d}	f_{2n+2-d}	...	f_{2n+1}
*	e_1	0	...	-1	-1	...	-1	...	-1	...	ν	...	ν	...	ν	ν	...	ν
⋮	⋮	⋮																
*	e_{c+1}	1	...	0	-1	...	-1	...	-1	...	ν	...	ν	...	ν	ν	...	ν
	e_{c+2}	1	...	1	h	...	-1	...	-1	...	-1	...	ν	...	ν	ν	...	ν
	⋮	⋮																
*	e_k	1	...	1	1	...	1	...	-1	...	-1	...	ν	...	ν	ν	...	ν
	⋮	⋮																
*	e_{n+1-d}	1	...	1	1	...	1	...	0	...	-1	...	-1	...	-1	ν	...	ν
⋮	⋮	⋮																
*	e_{n+c+2}	$-\nu$	...	$-\nu$	1	...	1	...	1	...	-1	...	-1	...	-1	-1	...	-1
	⋮	⋮																
*	e_{n+k+1}	$-\nu$	...	$-\nu$	$-\nu$	...	$-\nu$	...	1	...	1	...	m	...	-1	-1	...	-1
	⋮	⋮																
	e_{2n+1-d}	$-\nu$	...	$-\nu$	$-\nu$	...	$-\nu$	...	1	...	1	...	1	...	-1	-1	...	-1
*	e_{2n+2-d}	$-\nu$	...	$-\nu$	$-\nu$	...	$-\nu$	...	$-\nu$	...	1	...	1	...	1	0	...	-1
⋮	⋮	⋮																
*	e_{2n+1}	$-\nu$	...	$-\nu$	$-\nu$	...	$-\nu$	...	$-\nu$	...	1	...	1	...	1	1	...	0

Figure 5. Payoff matrix for game of Theorem 8.5.

such e_i against f_j in W_2^1, where we have $j \le i \le j+n$. If $j < i$ then every $e_{i,j} = 1$, and if $j = i = c+2$ then $a_{i,j} \le 0$, so e_k dominates. For f_j in W_2^2 we have $i < j \le i+n$, so every $a_{i,j} = -1$, and for f_j in W_2^3, $j > i+n$ so every $a_{i,j} = \nu$. Thus e_k dominates in this group against all of W_2.

(ii) Let $n+c+3 \le i \le 2n+1-d$. For f_j in W_2^1 we have $i > j+n$, so that every $a_{i,j} = -\nu$, and for f_j in W_2^2 we have $j < i \le j+n$, so that every $a_{i,j} = 1$. For f_j in W_2^3 we have $i \le j \le i+n$. If $i < j$ then $a_{i,j} = -1$, and if $i = j = 2n+1-d$ then $a_{i,j} = z = -1$ by hypothesis. Thus all e_i in this group are equivalent against W_2.

To complete the proof we show that against W_1 every f_j in $\tilde{W}_2 \setminus W_2$ is dominated by one in W_2, as follows:

(i) f_{c+2} dominates f_j for $c+2 \le j \le k$,

(ii) f_{n+1-d} dominates f_j for $k+1 \le j \le n+1-d$,

(iii) f_{n+c+2} dominates f_j for $n+c+2 \le j \le n+k$, and

(iv) f_{2n+1-d} dominates f_j for $n+k+1 \le j \le 2n+1-d$.

For (i), let $c+2 \le j \le k$. For all $i \le c+1$ we have $a_{i,j} = -1$. For $k \le i \le n+c+2$ we have $j \le i \le j+n$. If $j < i$ then $a_{i,j} = 1$, and if $j = i = k$ then $a_{i,j} = 1$ also, by hypothesis. For the remaining e_i in W_1 we have $i > j+n$ so that every $a_{i,j} = -\nu$. Thus the f_j in this group are equivalent against all e_i in W_1.

(ii) Let $k+1 \le j \le n+1-d$. For e_i in W_1^1 we have $i < j \le i+n$, so every $a_{i,j} = -1$. For e_i in W_1^2, $j \le i \le j+n$. If $j < i$ then each $a_{i,j} = 1$, and if $i = j = n+1-d$ then $a_{i,j} = 0$, so f_{n+1-d} dominates. For e_i in W_1^3, $i > j+n$ and every $a_{i,j} = -\nu$. Thus f_{n+1-d} dominates the f_j in this group against all of W_1.

(iii) Let $n+c+2 \le j \le n+k$. For e_i with $i \le c+1$, every $a_{i,j} = +\nu$. For $k \le i \le n+c+2$ we have $i \le j \le i+n$. If $i < j$ then every $a_{i,j} = -1$, and if $i = j = n+c+2$ then $a_{i,j} = y = -1$ as well. For the remaining e_i in W_1 we have $j < i \le j+n$, so that every $a_{i,j} = 1$. Thus all f_j in this group are equivalent against W_1.

(iv) Let $n+k+1 \le j \le 2n+1-d$. For e_i in W_1^1 we have $j > i+n$, so every $a_{i,j} = \nu$. For e_i in W_1^2, $i \le j \le i+n$. If $i < j$ then $a_{i,j} = -1$, and if $i = j = n+k+1$, then $a_{i,j} \ge -1$, so f_{2n+1-d} dominates. For e_i in W_1^3, $j < i \le j+n$, so every $a_{i,j} = 1$. Thus f_{2n+1-d} dominates the f_j in this group against all of W_1, and the proof is complete. □

The cases (ivBH) and (ix) are covered in the next theorem. For (ix) we formally regard $a = b = n$. If in (ivB) both G and H include a +1, both Theorems 8.5 and 8.6 apply, giving different but isomorphic reduced games.

THEOREM 8.6. Assume that $a > c$, $b \geq d$, $w = x = y = z = -1$, and that for some k with $c+4 \leq k \leq n-d-2$, $+1$ occurs on the diagonal in position $n+k$. Let

$W_1^1 = \{e_i\colon 1 \leq i \leq c+1\} \cup \{e_k\}$,

$W_1^2 = \{e_i\colon n+1-d \leq i \leq n+c+2\} \cup \{e_{n+k}\}$,

$W_1^3 = \{e_i\colon 2n+2-d \leq i \leq 2n+1\}$,

$W_2^1 = \{f_j\colon 1 \leq j \leq c+2\}$,

$W_2^2 = \{f_j\colon n+1-d \leq j \leq n+c+2\}$,

$W_2^3 = \{f_j\colon 2n+1-d \leq j \leq 2n+1\}$,

and $W_i = W_i^1 \cup W_i^2 \cup W_i^3$ for $i = 1,2$. Then mixed strategies which are optimal for the $(2c+2d+5)$ by $(2c+2d+5)$ subgame on $W_1 \times W_2$ are optimal for the full game on $\tilde{W}_1 \times \tilde{W}_2$. The reduced game is the balanced game with diagonal (8.0.5B').

PROOF. The game matrix is shown in Figure 6. We show first that against W_2, every e_i in $\tilde{W}_1 \setminus W_1$ is dominated by one in W_1, as follows:

(i) e_k dominates e_i for $c+2 \leq i \leq n-d$, and

(ii) e_{n+k} dominates e_i for $n+c+3 \leq i \leq 2n+1-d$.

For (i), let $c+2 \leq i \leq n-d$, and consider first such e_i against f_j in W_2^1, where we have $j \leq i \leq j+n$. When $j < i$ each $a_{i,j} = 1$, and for $j = i = c+2$, $a_{i,j} \leq 1$, so e_k dominates. For f_j in W_2^2 every $a_{i,j} = -1$ and for

		*	...	*	*				*	...	*				*	*	...	*
		f_1	...	f_{c+1}	f_{c+2}	...	f_k	...	f_{n+1-d}	...	f_{n+c+2}	...	f_{n+k}	...	f_{2n+1-d}	f_{2n+2-d}	...	f_{2n+1}
*	e_1	0	...	−1	−1	...	−1	...	−1	...	ν	...	ν	...	ν	ν	...	ν
⋮	⋮	⋮																
*	e_{c+1}	1	...	0	−1	...	−1	...	−1	...	ν	...	ν	...	ν	ν	...	ν
	e_{c+2}	1	...	1	h	...	−1	...	−1	...	−1	...	ν	...	ν	ν	...	ν
	⋮	⋮																
*	e_k	1	...	1	1	...	m	...	−1	...	−1	...	−1	...	ν	ν	...	ν
	⋮	⋮																
*	e_{n+1-d}	1	...	1	1	...	1	...	0	...	−1	...	−1	...	−1	ν	...	ν
⋮	⋮	⋮																
*	e_{n+c+2}	$-\nu$	...	$-\nu$	1	...	1	...	1	...	−1	...	−1	...	−1	−1	...	−1
	⋮	⋮																
*	e_{n+k}	$-\nu$	...	$-\nu$	$-\nu$	...	1	...	1	...	1	...	1	...	−1	−1	...	−1
	⋮	⋮																
	e_{2n+1-d}	$-\nu$	...	$-\nu$	$-\nu$	...	$-\nu$	...	1	...	1	...	1	...	−1	−1	...	−1
*	e_{2n+2-d}	$-\nu$	...	$-\nu$	$-\nu$	...	$-\nu$	...	$-\nu$	...	1	...	1	...	1	0	...	−1
⋮	⋮	⋮																
*	e_{2n+1}	$-\nu$	...	$-\nu$	$-\nu$	...	$-\nu$	...	$-\nu$	...	1	...	1	...	1	1	...	0

Figure 6. Payoff matrix for game of Theorem 8.6.

f_j in W_2^3 every $a_{i,j} = \nu$, so e_k dominates these e_i against all of W_2.

(ii) Let $n+c+3 \le i \le 2n+1-d$. For f_j in W_2^1 every $a_{i,j} = -\nu$, and for f_j in W_2^2 every $a_{i,j} = 1$. For f_j in W_2^3 we have $i \le j \le i+n$. If $i < j$ then $a_{i,j} = -1$, and if $i = j = 2n+1-d$ then $a_{i,j} = z = -1$ also. Thus the e_i in this group are equivalent against W_2.

To complete the proof we show that against W_1 every f_j in $\widetilde{W}_2 \setminus W_2$ is dominated by one in W_2, as follows.

(i) f_{c+2} dominates f_j for $c+2 \le j \le k-1$,

(ii) f_{n+1-d} dominates f_j for $k \le j \le n+1-d$,

(iii) f_{n+c+2} dominates f_j for $n+c+2 \le j \le n+k$, and

(iv) f_{2n+1-d} dominates f_j for $n+k+1 \le j \le 2n+1-d$.

For (i), let $c+2 \le j \le k-1$. For $1 \le i \le c+1$ we have every $a_{i,j} = -1$ and for $k \le i \le n+c+2$ every $a_{i,j} = 1$. For $i \ge n+k$ every $a_{i,j} = -\nu$, so the f_j in this group are equivalent against W_1.

(ii) Let $k \le j \le n+1-d$.

For e_i in W_1^1 we have $i \le j \le i+n$. If $i < j$ then every $a_{i,j} = -1$, and if $i = j = k$ then $a_{i,j} \ge -1$, so f_{n+1-d} dominates. For e_i in W_1^2 we have $j \le i \le j+n$. If $j < i$ then every $a_{i,j} = 1$, and if $j = i = n+1-d$ then $a_{i,j} = 0$, so f_{n+1-d} dominates. For e_i in W_1^3, $i > j+n$ so that every $a_{i,j} = -\nu$. Thus f_{n+1-d} dominates the f_j in this group against all e_i in W_1.

(iii) Let $n+c+2 \le j \le n+k$. For e_i with $i \le c+1$ every $a_{i,j} = \nu$. For $k \le i \le n+c+2$ we have $i \le j \le i+n$. If $i < j$ every $a_{i,j} = -1$, and if $i = j = n+c+2$ then $a_{i,j} = y = -1$ also. For the remaining e_i in W_1 we have $j \le i \le j+n$. If $j < i$ then every $a_{i,j} = 1$, and if $j = i = n+k$ then $a_{i,j} = 1$ by hypothesis. Thus the f_j in this group are equivalent against W_1.

(iv) Let $n+k+1 \le j \le 2n+1-d$. For e_i in W_1^1 every $a_{i,j} = \nu$, and for e_i in W_1^2 every $a_{i,j} = -1$. For e_i in W_1^3 every $a_{i,j} = 1$, so the f_j in this group are likewise equivalent against W_1, and the proof is complete. □

Next we deal with the cases (ivCG) and (xi).

THEOREM 8.7. Assume that $a \le c$, $b < d$, $w = x = y = z = -1$, and that for some k with $a+3 \le k \le n-b-2$, $+1$ occurs on the diagonal in position k. Let

$$W_1^1 = \{e_i\colon 1 \le i \le a+1\} \cup \{a_k\},$$

$$W_1^2 = \{e_i\colon n+1-b \le i \le n+a+1\} \cup \{a_{n+k+1}\},$$

$$W_1^3 = \{e_i\colon 2n+1-b \le i \le 2n+1\},$$

$W_2^1 = \{f_j\colon 1 \le j \le a+1\}$,

$W_2^2 = \{f_j\colon n-b \le j \le n+a+2\}$,

$W_2^3 = \{f_j\colon 2n+1-b \le j \le 2n+1\}$,

and $W_i = W_i^1 \cup W_i^2 \cup W_i^3$ for $i = 1,2$. Then mixed strategies which are optimal for the $(2a+2b+5)$ by $(2a+2b+5)$ subgame on $W_1 \times W_2$ are optimal for the full game on $\widetilde{W}_1 \times \widetilde{W}_2$. The reduced game is the balanced game with diagonal (8.0.5C).

PROOF. The matrix is shown in Figure 7. We show first that against W_2, every element of $\widetilde{W}_1 \setminus W_1$ is dominated by one in W_1, as follows:

(i) e_k dominates e_i for $a+2 \le i \le n-b$, and

(ii) e_{n+k+1} dominates e_i for $n+a+2 \le i \le 2n-b$.

For (i) let $a+2 \le i \le n-b$ and consider first such e_i against f_j in W_2^1. Then $j < i \le j+n$ so every $a_{i,j} = 1$. For f_j in W_2^2 we have $i \le j \le i+n$. If $i < j$ every $a_{i,j} = -1$, and if $i = j = n-b$ then $a_{i,j} = x = -1$ also. For f_j in W_2^3 we have $j > i+n$ so every $a_{i,j} = \nu$. Thus the e_i in this group are equivalent against W_2.

(ii) Let $n+a+2 \le i \le 2n-b$. For f_j in W_2^1 we have $i > j+n$, so every $a_{i,j} = -\nu$. For f_j in W_2^2 we have $j \le i \le j+n$. If $j < i$ then every $a_{i,j} = 1$, and if $j = i = n+a+2$ then $a_{i,j} \le 0$, so e_{n+k+1} dominates. For

		*	...	*				*	*	...	*	*				*	...	*
		f_1	...	f_{a+1}	...	f_k	...	f_{n-b}	f_{n+1-b}	...	f_{n+a+1}	f_{n+a+2}	...	f_{n+k+1}	...	f_{2n+1-b}	...	f_{2n+1}
*	e_1	0	...	−1	...	−1	...	−1	−1	...	ν	ν	...	ν	...	ν	...	ν
$\vdots$	$\vdots$	$\vdots$																
*	e_{a+1}	1	...	−1	...	−1	...	−1	−1	...	−1	ν	...	ν	...	ν	...	ν
	$\vdots$	$\vdots$																
*	e_k	1	...	1	...	1	...	−1	−1	...	−1	−1	...	ν	...	ν	...	ν
	$\vdots$	$\vdots$																
	e_{n-b}	1	...	1	...	1	...	−1	−1	...	−1	−1	...	−1	...	ν	...	ν
*	e_{n+1-b}	1	...	1	...	1	...	1	0	...	−1	−1	...	−1	...	−1	...	ν
$\vdots$	$\vdots$	$\vdots$																
*	e_{n+a+1}	$-\nu$	...	1	...	1	...	1	1	...	0	−1	...	−1	...	−1	...	−1
	e_{n+a+2}	$-\nu$	...	$-\nu$	...	1	...	1	1	...	1	h	...	−1	...	−1	...	−1
	$\vdots$	$\vdots$																
*	e_{n+k+1}	$-\nu$	...	$-\nu$	...	$-\nu$	...	1	1	...	1	1	...	m	...	−1	...	−1
	$\vdots$	$\vdots$																
*	e_{2n+1-b}	$-\nu$	...	$-\nu$	...	$-\nu$	...	$-\nu$	1	...	1	1	...	1	...	0	...	−1
$\vdots$	$\vdots$	$\vdots$																
*	e_{2n+1}	$-\nu$	...	$-\nu$	...	$-\nu$	...	$-\nu$	$-\nu$	...	1	1	...	1	...	1	...	0

Figure 7. Game matrix for Theorem 8.7.

f_j in W_2^3 we have $i < j \le i+n$ so that every $a_{i,j} = -1$. Thus e_{n+k+1} dominates in this group of e_i against all f_j in W_2.

To complete the proof we show that against W_1, every f_j in $\tilde{W}_2 \setminus W_2$ is dominated by one in W_2, as follows:

(i) f_{a+1} dominates f_j for $a+1 \le j \le k$,

(ii) f_{n-b} dominates f_j for $k+1 \le j \le n-b$,

(iii) f_{n+a+2} dominates f_j for $n+a+2 \le j \le n+k$, and

(iv) f_{2n+1-b} dominates f_j for $n+k+1 \le j \le 2n+1-b$.

For (i), let $a+1 \le j \le k$ and consider first such f_j against e_i with $1 \le i \le a+1$, where we have $i \le j \le i+n$. If $i < j$ every $a_{i,j} = -1$, and if $i = j = a+1$ then $a_{i,j} = w = -1$ also. Next consider such f_j against e_i with $k \le i \le n+a+1$, where we have $j \le i \le j+n$. If $j < i$ then $a_{i,j} = 1$, and if $j = i = k$ then $a_{i,j} = 1$ by hypothesis. Finally, for e_i with $i \ge n+k+1$ all $a_{i,j} = -\nu$. Thus the f_j in this group are equivalent against W_1.

(ii) Let $k+1 \le j \le n-b$. For e_i in W_1^1 we have $i < j \le i+n$, so every $a_{i,j} = -1$. For e_i in W_1^2, $j < i \le j+n$ and every $a_{i,j} = 1$. For e_i in W_1^3, $i > j+n$ so every $a_{i,j} = -\nu$. Thus all f_j in this group are equivalent against W_1.

(iii) Let $n+a+2 \le j \le n+k$. For e_i with $1 \le i \le a+1$, we have $j > i+n$ so every $a_{i,j} = \nu$. For e_i with $k \le i \le n+a+1$, $i < j \le i+n$ and every $a_{i,j} = -1$. For the remaining e_i in W_1 we have $j < i \le j+n$ so that every $a_{i,j} = 1$. Thus the f_j in this group too are equivalent against all of W_1.

(iv) Let $n+k+1 \le j \le 2n+1-b$. For e_i in W_1^1 every $a_{i,j} = \nu$. For e_i in W_1^2 we have $i \le j \le i+n$. If $i < j$ then $a_{i,j} = -1$, and if $i = j = n+k+1$, then $a_{i,j} \ge -1$, so f_{2n+1-b} dominates. For e_i in W_1^3 we have $j \le i \le j+n$. If $j < i$ then $a_{i,j} = 1$, and if $j = i = 2n+1-b$ then $a_{i,j} = 0$, so f_{2n+1-b} dominates. Thus f_{2n+1-b} dominates the f_j in this group against all e_i in W_1, and the proof is complete. □

The remaining subcase which reduces to a game of odd order is ivC with + on the right.

THEOREM 8.8. Assume that $a \le c$, $b < d$, $w = x = y = z = -1$, and that for some k with $c+4 \le k \le n-d-1$, +1 occurs on the diagonal in position $n+k$. Let

$W_1^1 = \{e_i\colon 1 \le i \le a+1\} \cup \{a_k\}$,

$W_1^2 = \{e_i\colon n+1-b \le i \le n+a+1\} \cup \{a_{n+k}\}$

$W_1^3 = \{e_i\colon 2n+1-b \le i \le 2n+1\}$,

$W_2^1 = \{f_j\colon 1 \le j \le a+1\}$,

$W_2^2 = \{f_j\colon n-b \le j \le n+a+2\}$,

$W_2^3 = \{f_j\colon 2n+1-b \le i \le 2n+1\}$,

and $W_i = W_i^1 \cup W_i^2 \cup W_i^3$ for $i = 1,2$. Then mixed strategies which are optimal for the (2a+2b+5) by (2a+2b+5) subgame on $W_1 \times W_2$ are optimal for the full game on $\hat{W}_1 \times \hat{W}_2$. The reduced game is the balanced game with diagonal (8.0.5C).

PROOF. The matrix is shown in Figure 8. We show first that against W_2 each e_i in $\hat{W}_1 \setminus W_1$ is dominated by an element of W_1, as follows:

(i) e_k dominates e_i for $a+2 \le i \le n-b$, and

(ii) e_{n+k} dominates e_i for $n+a+2 \le i \le 2n-b$.

For (i), let $a+2 \le i \le n-b$, and consider first such e_i against f_j in W_2^1. Then $j < i \le j+n$, and therefore every $a_{i,j} = 1$. For f_j in W_2^2 we have $i \le j \le i+n$. If $i < j$ then $a_{i,j} = -1$, and if $i = j = n-b$ then $a_{i,j} = x = -1$ also. For f_j in W_2^3 we have $j > i+n$ and hence every $a_{i,j} = \nu$. Thus the e_i in this group are equivalent against W_2.

(ii) Let $n+a+2 \le i \le 2n-b$. For f_j in W_2^1 we have $i > j+n$, so every $a_{i,j} = -\nu$. For f_j in W_2^2 we have $j \le i \le j+n$. If $j < i$ then $a_{i,j} = 1$, and if $j = i =$

		*	...	*				*	*	...	*	*				*	...	*
		f_1	...	f_{a+1}	...	f_k	...	f_{n-b}	f_{n+1-b}	...	f_{n+a+1}	f_{n+a+2}	...	f_{n+k}	...	f_{2n+1-b}	...	f_{2n+1}
*	e_1	0	...	−1	...	−1	...	−1	−1	...	ν	ν	...	ν	...	ν	...	ν
⋮	⋮	⋮																
*	e_{a+1}	1	...	−1	...	−1	...	−1	−1	...	−1	ν	...	ν	...	ν	...	ν
	⋮	⋮																
*	e_k	1	...	1	...	h	...	−1	−1	...	−1	−1	...	−1	...	ν	...	ν
	⋮	⋮																
	e_{n-b}	1	...	1	...	1	...	−1	−1	...	−1	−1	...	−1	...	ν	...	ν
*	e_{n+1-b}	1	...	1	...	1	...	1	0	...	−1	−1	...	−1	...	−1	...	ν
⋮	⋮	⋮																
*	e_{n+a+1}	$-\nu$	...	1	...	1	...	1	1	...	0	−1	...	−1	...	−1	...	−1
	e_{n+a+2}	$-\nu$	...	$-\nu$	...	1	...	1	1	...	1	m	...	−1	...	−1	...	−1
	⋮	⋮																
*	e_{n+k}	$-\nu$	...	$-\nu$	...	1	...	1	1	...	1	1	...	1	...	−1	...	−1
	⋮	⋮																
*	e_{2n+1-b}	$-\nu$	...	$-\nu$	...	$-\nu$	...	$-\nu$	1	...	1	1	...	1	...	0	...	−1
⋮	⋮	⋮																
*	e_{2n+1}	$-\nu$	...	$-\nu$	...	$-\nu$	...	$-\nu$	$-\nu$	...	1	1	...	1	...	1	...	0

Figure 8. Game matrix for Theorem 8.8.

$n+a+2$ then $a_{i,j} \le 1$, so e_{n+k} dominates. For f_j in W_2^3 we have $i < j \le i+n$, whence every $a_{i,j} = -1$. Thus e_{n+k} dominates the e_i in this group against all of W_2.

To complete the proof we show that against W_1 each f_j in $\tilde{W}_2 \setminus W_2$ is dominated by one in W_2, as follows:

(i) f_{a+1} dominates f_j for $a+1 \le j \le k-1$,

(ii) f_{n-b} dominates f_j for $k \le j \le n-b$,

(iii) f_{n+a+2} dominates f_j for $n+a+2 \le j \le n+k$, and

(iv) f_{2n+1-b} dominates f_j for $n+k+1 \le j \le 2n+1-b$.

For (i), let $a+1 \le j \le k-1$, and consider first such f_j against e_i with $i \le a+1$. If $i < a+1$ then $i < j \le i+n$, and every $a_{i,j} = -1$. If $i = j = a+1$ then $a_{i,j} = w = -1$ also. Next consider such f_j against e_i with $k \le i \le n+a+1$. Then $j < i \le j+n$, so every $a_{i,j} = 1$. For the remaining e_i in W_1 we have $i > j+n$ so that every $a_{i,j} = -\nu$. Thus the f_j in this group are equivalent against all of W_1.

(ii) Let $k \le j \le n-b$. For e_i in W_1^1 we have $i \le j \le i+n$. If $i < j$ then $a_{i,j} = -1$, and if $i = j = k$, then $a_{i,j} \ge -1$, so f_{n-b} dominates. For e_i in W_1^2 we have $j < i \le j+n$, and every $a_{i,j} = 1$. For e_i in W_1^3, $i > j+n$ so every $a_{i,j} = -\nu$. Thus f_{n-b} dominates in this group against all W_1.

(iii) Let $n+a+2 \le j \le n+k$. Then for e_i with $i \le a+1$, every $a_{i,j} = \nu$. For e_i with $k \le i \le n+a+1$ we have $i < j \le i+n$, so that every $a_{i,j} = -1$. For the remaining e_i in W_1, $j \le i \le j+n$. If $j < i$ then $a_{i,j} = 1$, and if $j = i = n+k$ then $a_{i,j} = 1$ by hypothesis. Thus the f_j in this group are equivalent against W_1.

(iv) Let $n+k+1 \le j \le 2n+1-b$. For e_i in W_1^1 we have $j > i+n$ so every $a_{i,j} = \nu$. For e_i in W_1^2, $i < j \le i+n$, and every $a_{i,j} = -1$. For e_i in W_1^3 we have $j \le i \le j+n$. If $j < i$ then $a_{i,j} = 1$, and if $j = i = 2n+1-b$ then $a_{i,j} = 0$, so f_{2n+1-b} dominates. Thus f_{2n+1-b} dominates the f_j in this group against all of W_1, and the proof is complete. □

9. Reduction of balanced games to even order.

In this section we describe the reduction of the remaining eighteen of the 36 cases in (8.0.3), (8.0.4) and (8.0.7). There are again four types of reduced game, corresponding to (A), (B), (C) and (D) in (8.0.4). In our description of these, the first nonzero main-diagonal element is again always -1, and off-diagonal zeros are concentrated in a middle segment of the first subdiagonal. The remainder of the matrix is the same in all cases, and may be described by the diagram in Figure 9.

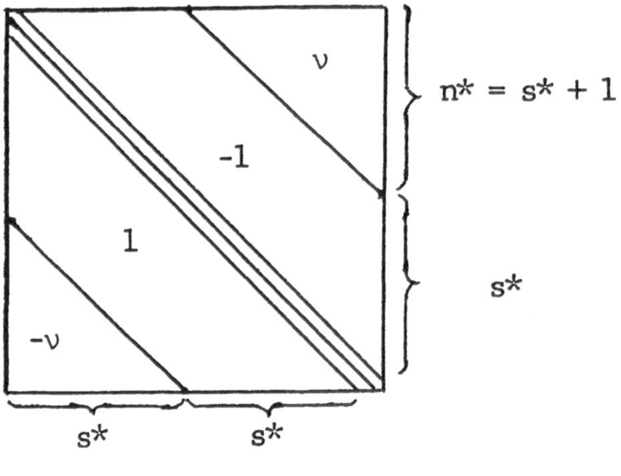

Figure 9.

If the order of the reduced game is $2n^*$, then each element of the n^* by n^* triangle in the upper right corner is ν, and each element in the s^* by s^* triangle in the lower left corner is $-\nu$. (Here $s^* = n^*-1$.) Between the main diagonal and the upper right triangle are s^* diagonals, each element of which is -1, and between the first subdiagonal and the lower left triangle are s^* diagonals, each element of which is 1. The four patterns on the main diagonal and first subdiagonal are

(9.0.1A) $$\begin{matrix} 0^a & & (-1)^{a+d+4} & & 0^d & \\ & 1^{a+1} & & 0^{a+d+1} & & 1^{d+1} \end{matrix}$$

(9.0.1B) $$\begin{matrix} 0^{c+1} & & (-1)^{c+d+3} & & 0^d & \\ & 1^{c+1} & & 0^{c+d+1} & & 1^{d+1} \end{matrix}$$

(9.0.1C) $$\begin{matrix} 0^a & & (-1)^{a+b+3} & & 0^{b+1} & \\ & 1^{a+1} & & 0^{a+b+1} & & 1^{b+1} \end{matrix}$$

(9.0.1D) $$\begin{matrix} 0^{c+1} & & (-1)^{b+c+4} & & 0^{b+1} & \\ & 1^{c+2} & & 0^{b+c+1} & & 1^{b+2} \end{matrix}$$

Our first theorem here deals with cases (iiB), (viB), (iiiB), (viiB) and (x). The theorem does not assume w = -1, and actually applies directly to (iiiB)' and (viiB)', where the sign sequences are opposite to those in (iii) and (vii). Cases (iiiB) and (viiB) are obtained then by interchanging the roles of the players.

THEOREM 9.1. Assume that $y = 1$, $z = -1$, $a > c$ and $b \geq d$. (We do not assume that $w = -1$.) Let

$W_1^1 = \{e_i: 1 \leq i \leq c+2\}$,

$W_1^2 = \{e_i\colon n+1-d \le i \le n+c+2\}$,

$W_1^3 = \{e_i\colon 2n+2-d \le i \le 2n+1\}$,

$W_2^1 = \{f_j\colon 1 \le j \le c+1\}$,

$W_2^2 = \{f_j\colon n+1-d \le j \le n+c+2\}$,

$W_2^3 = \{f_j\colon 2n+1-d \le j \le 2n+1\}$,

and $W_i = W_i^1 \cup W_i^2 \cup W_i^3$ for $i = 1,2$. Then optimal strategies for the $(2c+2d+4)$ by $(2c+2d+4)$ subgame on $W_1 \times W_2$ are optimal for the full game on $\tilde{W}_1 \times \tilde{W}_2$. The reduced game is of type (9.0.1B).

PROOF. We show first that against W_2, each element of $\tilde{W}_1 \setminus W_1$ is dominated by an element of W_1, as follows:

(i) e_{c+2} dominates e_i for $c+2 \le i \le n-d$, and

(ii) e_{n+c+2} dominates e_i for $n+c+2 \le i \le 2n+1-d$.

(See Figure 10 for the payoff matrix of the game.)

For (i), let $c+2 \le i \le n-d$, and consider first such e_i against f_j in W_2^1. Since $j \le c+1 < i < n+j$, every $a_{i,j} = 1$. Next consider such e_i against f_j in W_2^2. Now $i < n+1-d \le j \le n+c+2 \le i+n$, and every $a_{i,j} = -1$. For f_j in W_2^3 we have $j > n+i$, so that every $a_{i,j} = \nu$. Thus, against W_2 all e_i in this group are in fact equivalent.

		*	...	*			*	...	*		*	*	...	*	
		f_1	...	f_{c+1}	f_{c+2}		f_{n+1-d}	...	f_{n+c+2}	...	f_{2n+1-d}	f_{2n+2-d}	...	f_{2n+1}	
*	e_1	0	...	−1	−1	...	−1	...	ν	...	ν	ν	...	ν	
⋮	⋮	⋮													
*	e_{c+1}	1	...	0	−1	...	−1	...	ν	...	ν	ν	...	ν	
*	e_{c+2}	1	...	1	h	...	−1	...	−1	...	ν	ν	...	ν	
	⋮	⋮													
*	e_{n+1-d}	1	...	1	1	...	0	...	−1	...	−1	ν	...	ν	
⋮	⋮	⋮													
*	e_{n+c+2}	$-\nu$	...	$-\nu$	1	...	1	...	y	...	−1	−1	...	−1	(y=1)
	⋮	⋮													
	e_{2n+1-d}	$-\nu$	...	$-\nu$	$-\nu$	...	1	...	1	...	z	−1	...	−1	(z=−1)
*	e_{2n+2-d}	$-\nu$	...	$-\nu$	$-\nu$	...	$-\nu$	...	1	...	1	0	...	−1	
⋮	⋮	⋮													
*	e_{2n+1}	$-\nu$	...	$-\nu$	$-\nu$	...	$-\nu$	...	1	...	1	1	...	0	

Figure 10. Matrix for game in Theorem 9.1.

(ii) Let $n+c+2 \le i \le 2n+1-d$, and consider first such e_i against f_j in W_2^1. Here $j \le c+1$, so $i > n+j$ and every $a_{i,j} = -\nu$. Next consider such e_i against f_j in W_2^2. Then $j \le i \le j+n$. If $j < i$ then $a_{i,j} = 1$, and if $j = i = n+c+2$ then $a_{i,j} = y = 1$ by hypothesis. Last, consider such e_i against f_j in W_2^3, where we have $i \le j < i+n$. If $i < j$ then $a_{i,j} = -1$, and if $i = j = 2n+1-d$, then $a_{i,j} = z = -1$ by hypothesis. Thus, against W_2 all e_i in this group are equivalent.

We complete the proof by showing that against W_1, each element of $\tilde{W}_2 \setminus W_2$ is dominated by an element of W_2, as follows:

(i) f_{n+1-d} dominates f_j for $c+2 \le j \le n+1-d$, and

(ii) f_{2n+1-d} dominates f_j for $n+c+3 \le j \le 2n+1-d$.

For (i), let $c+2 \le j \le n+1-d$, and consider first such f_j against e_i in W_1^1. Then $i \le j \le i+n$. If $i < j$ then $a_{i,j} = -1$, and if $i = j = c+2$, then $a_{i,j} \ge -1$, so f_{n+1-d} dominates. Next consider such f_j against e_i in W_1^2. Then $j \le i \le j+n$. If $j < i$ we have $a_{i,j} = 1$, and if $j = i = n+1-d$, then $a_{i,j} = 0$ since $b \ge d$. Thus $a_{i,n+1-d} \le a_{i,j}$ in each case. Last, consider such f_j against e_i in W_1^3. Then $i > j+n$ so that every $a_{i,j} = -\nu$.

Thus f_{n+1-d} dominates the other f_j in this group against all of W_1.

(ii) Let $n+c+3 \le j \le 2n+1-d$, and consider first such f_j against e_i in W_1^1. Then $j > n+i$, and each $a_{i,j} = \nu$. For e_i in W_1^2 we have $i < j \le i+n$, so that each $a_{i,j} = -1$. Finally, for e_i in W_1^3 we have $j < i < j+n$ and every $a_{i,j} = 1$. Thus all f_j in this group are equivalent against W_1, and the proof is complete. □

The next theorem deals with cases (vC), (viC), (viiC), (viiiC) and (xii).

THEOREM 9.2. Assume that $w = -1$, $x = 1$, $a \le c$ and $b < d$. Let

$$W_1^1 = \{e_i\colon 1 \le i \le a+1\},$$

$$W_1^2 = \{e_i\colon n-b \le i \le n+a+1\},$$

$$W_1^3 = \{e_i\colon 2n+1-b \le i \le 2n+1\},$$

$$W_2^1 = \{f_j\colon 1 \le j \le a+1\},$$

$$W_2^2 = \{f_j\colon n+1-b \le j \le n+a+2\},$$

$$W_2^3 = \{f_j\colon 2n+1-b \le j \le 2n+1\},$$

and $W_i = W_i^1 \cup W_i^2 \cup W_i^3$, $i = 1,2$. Then optimal strategies for the $(2a+2b+4)$ by $(2a+2b+4)$ subgame on $W_1 \times W_2$ are optimal for the full game on $\hat{W}_1 \times \hat{W}_2$. The reduced game is of type (9.0.1C).

PROOF. We show first that against W_2, each element of $\widetilde{W}_1 \setminus W_1$ is dominated by an element of W_1, as follows:

(i) e_{n-b} dominates e_i for $a+2 \leq i \leq n-b$, and

(ii) e_{2n+1-b} dominates e_i for $n+a+2 \leq i \leq 2n+1-b$.

(See Figure 11 for the payoff matrix.)

For (i), let $a+2 \leq i \leq n-b$, and consider first such e_i against f_j in W_2^1. Since $j \leq a+1$ we have $j < i < j+n$, and every $a_{i,j} = 1$. For f_j in W_2^2, $i < j \leq i+n$, so that every $a_{i,j} = -1$, and for f_j in W_2^3, $j > i+n$ and therefore every $a_{i,j} = \nu$. Thus, against W_2 these e_i are equivalent.

(ii) Let $n+a+2 \leq i \leq 2n+1-b$, and consider first such e_i against f_j in W_2^1. Since $i > j+n$, every $a_{i,j} = -\nu$. For f_j in W_2^2 we have $j \leq i \leq j+n$. If $j < i$ then $a_{i,j} = 1$, and if $j = i = n+a+2$ then $a_{i,j} \leq 1$, so against W_2^2, e_{2n+1-b} dominates the e_i in this group. For f_j in W_2^3 we have $i \leq j \leq i+n$. If $i < j$ each $a_{i,j} = -1$, and if $i = j = 2n+1-b$ then $a_{i,j} = 0$. Thus e_{2n+1-b} dominates the e_i in this group against all f_j in W_2.

		*	...	*			*	...	*	*		*	...	*	
		f_1	...	f_{a+1}	...	f_{n-b}	f_{n+1-b}	...	f_{n+a+1}	f_{n+a+2}	...	f_{2n+1-b}	...	f_{2n+1}	
*	e_1	0	...	-1	...	-1	-1	...	ν	ν	...	ν	...	ν	
⋮	⋮	⋮													
*	e_{a+1}	1	...	w	...	-1	-1	...	-1	ν	...	ν	...	ν	(w=-1)
	⋮	⋮													
*	e_{n-b}	1	...	1	...	x	-1	...	-1	-1	...	ν	...	ν	(x=1)
*	e_{n+1-b}	1	...	1	...	1	0	...	-1	-1	...	-1	...	ν	
⋮	⋮	⋮													
*	e_{n+a+1}	$-\nu$	...	1	...	1	1	...	0	-1	...	-1	...	-1	
	e_{n+a+2}	$-\nu$	...	$-\nu$	...	1	1	...	1	h	...	-1	...	-1	
	⋮	⋮													
*	e_{2n+1-b}	$-\nu$	...	$-\nu$	...	$-\nu$	1	...	1	1	...	0	...	-1	
⋮	⋮	⋮													
*	e_{2n+1}	$-\nu$	...	$-\nu$	...	$-\nu$	$-\nu$	...	1	1	...	1	...	0	

Figure 11. Matrix for game in Theorem 9.2.

To complete the proof we show that against W_1 each element of $\widetilde{W}_2 \setminus W_2$ is dominated by one in W_2, as follows:

(i) f_{a+1} dominates f_j for $a+1 \le j \le n-b$, and

(ii) f_{n+a+2} dominates f_j for $n+a+2 \le j \le 2n-b$.

For (i), let $a+1 \le j \le n-b$, and consider first such f_j against e_i in W_1^1, where we have $i \le j < i+n$. If $i < j$ each $a_{i,j} = -1$, and if $i = j = a+1$ then $a_{i,j} = w = -1$ by hypothesis, so, against W_1^1 all f_j in this group are equivalent. Next consider such f_j against e_i in W_1^2, where we have $j \le i \le j+n$. If $j < i$ then $a_{i,j} = 1$, and if $j = i = n-b$ then $a_{i,j} = x = 1$ by hypothesis, so against W_1^2 these f_j are again equivalent. For e_i in W_1^3, $i > j+n$, so every $a_{i,j} = -\nu$. Thus all f_j in this group are equivalent against W_1.

(ii) Let $n+a+2 \le j \le 2n-b$. For e_i in W_1^1 we have $j > i+n$, so that every $a_{i,j} = \nu$. For e_i in W_1^2, $i < j \le i+n$ and hence every $a_{i,j} = -1$. For e_i in W_1^3 we have $j < i < j+n$, and every $a_{i,j} = 1$. Thus all f_j in this group are equivalent against W_1, and the proof is complete. □

The next theorem handles cases (vD) and (viD).

THEOREM 9.3. Assume that $w = -1$, $x = y = 1$, $a > c$ and $b < d$. Let

$$W_1^1 = \{e_i: 1 \le i \le c+2\},$$

$$W_1^2 = \{e_i: n-b \le i \le n+c+2\},$$

$$W_1^3 = \{e_i: 2n+1-b \le i \le 2n+1\},$$

$$W_2^1 = \{f_j: 1 \le j \le c+1\} \cup \{f_{a+1}\},$$

$$W_2^2 = \{f_j: n+1-b \le j \le n+c+2\} \cup \{f_{n+a+2}\},$$

$$W_2^3 = \{f_j: 2n+1-b \le j \le 2n+1\},$$

and $W_i = W_i^1 \cup W_i^2 \cup W_i^3$ for $i = 1,2$. Then optimal strategies for the (2b+2c+6) by (2b+2c+6) subgame on $W_1 \times W_2$ are optimal for the full game on $\widetilde{W}_1 \times \widetilde{W}_2$. The reduced game is of type (9.0.1D).

PROOF. We show first that against W_2, each element of $\widetilde{W}_1 \setminus W_1$ is dominated by an element of W_1, as follows:

(i) e_{c+2} dominates e_i for $c+2 \le i \le a+1$,

(ii) e_{n-b} dominates e_i for $a+2 \le i \le n-b$,

(iii) e_{n+c+2} dominates e_i for $n+c+2 \le i \le n+a+1$,

and

(iv) e_{2n+1-b} dominates e_i for $n+a+2 \le i \le 2n+1-b$.

(See Figure 12 for the matrix of the game.)

For (i), let $c+2 \le i \le a+1$, and consider first such e_i against f_j with $j \le c+1$. Then $j < i < j+n$ and

		*	...	*			*			*	...	*				*	...	*	
		f_1	...	f_{c+1}	f_{c+2}	...	f_{a+1}	...	f_{n-b}	f_{n+1-b}	...	f_{n+c+2}	...	f_{n+a+2}	...	f_{2n+1-b}	...	f_{2n+1}	
*	e_1	0	...	−1	−1	...	−1	...	−1	−1	–...	ν	...	ν	...	ν	...	ν	
⋮	⋮	⋮																	
*	e_{c+1}	1	...	0	−1	...	−1	...	−1	−1	...	ν	...	ν	...	ν	...	ν	
*	e_{c+2}	1	...	1	0	...	−1	...	−1	−1	...	−1	...	ν	...	ν	...	ν	
	⋮	⋮																	
	e_{a+1}	1	...	1	1	...	w	...	−1	−1	...	−1	...	ν	...	ν	...	ν	(w=−1)
	⋮	⋮																	
*	e_{n-b}	1	...	1	1	...	1	...	x	−1	...	−1	...	−1	...	ν	...	ν	(x=1)
*	e_{n+1-b}	1	...	1	1	...	1	...	1	0	...	−1	...	−1	...	−1	...	ν	
	⋮	⋮																	
*	e_{n+c+2}	$-\nu$	...	$-\nu$	1	...	1	...	1	1	...	y	...	−1	...	−1	...	−1	(y=1)
	⋮	⋮																	
	e_{n+a+2}	$-\nu$	...	$-\nu$	$-\nu$	...	$-\nu$	...	1	1	...	1	...	h	...	−1	...	−1	
	⋮	⋮																	
*	e_{2n+1-b}	$-\nu$	...	$-\nu$	$-\nu$	...	$-\nu$	...	$-\nu$	1	...	1	...	1	...	0	...	−1	
⋮	⋮	⋮																	
*	e_{2n+1}	$-\nu$	...	$-\nu$	$-\nu$	...	$-\nu$	...	$-\nu$	$-\nu$	...	1	...	1	...	1	...	0	

Figure 12. Payoff matrix for the game of Theorem 9.3.

(If a = c+1, row c+2 and column c+2 should be deleted.)

every $a_{i,j} = 1$. Next consider such e_i against f_j with $a+1 \le j \le n+c+2$, where we have $i \le j \le i+n$. If $i < j$ then $a_{i,j} = -1$, and if $i = j = a+1$ then $a_{i,j} = w = -1$ also. Lastly consider such e_i against f_j with $n+a+2 \le j \le 2n+1$. Then $j > i+n$, so every $a_{i,j} = \nu$. Thus against W_2 all e_i in this group are equivalent.

(ii) Let $a+2 \le i \le n-b$. For f_j in W_2^1 we have $j < i < j+n$, and every $a_{i,j} = 1$. For f_j in W_2^2 we have $i < j \le i+n$ so that every $a_{i,j} = -1$, and for f_j in W_2^3, $j > i+n$ and every $a_{i,j} = \nu$. Thus all e_i in this group are equivalent against W_2.

(iii) Let $n+c+2 \le i \le n+a+1$. For $j \le c+1$ we have $i > n+j$ so every $a_{i,j} = -\nu$. For $a+1 \le j \le n+c+2$ we have $j \le i \le j+n$. If $j < i$ then $a_{i,j} = 1$ in every case, and if $j = i = n+c+2$ then $a_{i,j} = y = 1$. For $j \ge n+a+2$ we have $i < j < i+n$, and hence every $a_{i,j} = -1$. Thus all e_i in this group are equivalent against W_2.

(iv) Let $n+a+2 \le i \le 2n+1-b$. For f_j in W_2^1 we have $i > j+n$ so every $a_{i,j} = -\nu$. For f_j in W_2^2 we have $j \le i \le j+n$. If $j < i$ then each $a_{i,j} = 1$, and if $j = i = n+a+2$ then $a_{i,j} \le 1$, so e_{2n+1-b} dominates. For f_j in W_2^3 we have $i \le j < i+n$. If $i < j$ then $a_{i,j} = -1$, and if

$i = j = 2n+1-b$ then $a_{i,j} = 0$, so again e_{2n+1-b} dominates. Thus against all f_j in W_2, e_{2n+1-b} dominates the e_i in this group.

To complete the proof we show that against W_1, each element of $\widetilde{W}_2 \setminus W_2$ is dominated by an element of W_2, as follows:

(i) f_{a+1} dominates f_j for $c+2 \leq j \leq n-b$, and

(ii) f_{n+a+2} dominates f_j for $n+c+3 \leq j \leq 2n-b$.

For (i), let $c+2 \leq j \leq n-b$, and consider first such f_j against e_i in W_1^1, where $i \leq j \leq i+n$. If $i < j$ then every $a_{i,j} = -1$. If $i = j = c+2 < a+1$ then $a_{i,j} = 0$, and if $i = j = c+2 = a+1$ then $a_{i,j} = w = -1$. In every case, f_{a+1} dominates. Next consider such f_j against e_i in W_1^2, where $j \leq i \leq j+n$. If $j < i$ then $a_{i,j} = 1$, and if $j = i = n-b$, then $a_{i,j} = x = 1$, so all f_j in this group are equivalent against W_1^2. For e_i in W_1^3 we have $i > j+n$, so that every $a_{i,j} = -\nu$. Thus, against all e_i in W_1, f_{a+1} dominates the f_j in this group.

(ii) Let $n+c+3 \leq j \leq 2n-b$. For e_i in W_1^1 we have $j > n+i$, so every $a_{i,j} = \nu$. For e_i in W_1^2, $i < j \leq i+n$, and every $a_{i,j} = -1$. For e_i in W_1^3 we have $j < i$, and therefore every $a_{i,j} = 1$. Thus the f_j in this group are equivalent against all e_i in W_1, and the proof is complete. □

The next theorem takes care of cases (viA) and (viiiA).

THEOREM 9.4 Assume that $w = -1$, $x = 1$, $z = -1$, $a \le c$ and $b \ge d$. Let

$W_1^1 = \{e_i\colon 1 \le i \le a+1\}$,

$W_1^2 = \{e_{n-b}\} \cup \{e_i\colon n+1-d \le i \le n+a+1\}$,

$W_1^3 = \{e_{2n+1-b}\} \cup \{e_i\colon 2n+2-d \le i \le 2n+1\}$,

$W_2^1 = \{f_j\colon 1 \le j \le a+1\}$,

$W_2^2 = \{f_j\colon n+1-d \le j \le n+a+2\}$,

$W_2^3 = \{f_j\colon 2n+1-d \le j \le 2n+1\}$,

and $W_i = W_i^1 \cup W_i^2 \cup W_i^3$ for $i = 1,2$. Then optimal strategies for the (2a+2d+4) by (2a+2d+4) subgame on $W_1 \times W_2$ are optimal for the full game on $\widetilde{W}_1 \times \widetilde{W}_2$. The reduced game is of type (9.0.1A).

PROOF. We show first that against W_2, every element of $\widetilde{W}_1 \setminus W_1$ is dominated by an element of W_1, as follows:

(i) e_{n-b} dominates all e_i with $a+2 \le i \le n-d$, and

(ii) e_{2n+1-b} dominates all e_i with $n+a+2 \le i \le 2n+1-d$.

(See Figure 13 for the matrix of the game.)

For (i), let $a+2 \le i \le n-d$, and consider first such e_i against f_j in W_2^1. Then $j < i < j+n$, so that each $a_{i,j} = 1$. For f_j in W_2^2 we have $i < j \le i+n$, so each $a_{i,j} = -1$, and for f_j in W_2^3, $j > i+n$ and each $a_{i,j} = \nu$. Thus against W_2, all e_i in this group are equivalent.

(ii) Let $n+a+2 \le i \le 2n+1-d$. For f_j in W_2^1 we have $i > j+n$, so that every $a_{i,j} = -\nu$. For f_j in W_2^2, $j \le i \le j+n$. If $j < i$ then $a_{i,j} = 1$, and if $j = i = n+a+2$, then $a_{i,j} \le 1$, so e_{2n+1-b} dominates. For f_j in W_2^3 we have $i \le j < i+n$. If $i < j$ then $a_{i,j} = -1$, and if $i = j = 2n+1-d$ then $a_{i,j} = z = -1$. Thus e_{2n+1-b} dominates the e_i in this group against all f_j in W_2.

To complete the proof we show that against W_1, every element of $\widetilde{W}_2 \setminus W_2$ is dominated by an element of W_2, as follows:

(i) f_{a+1} dominates f_j for $a+1 \le j \le n-b$,

(ii) f_{n+1-d} dominates f_j for $n+1-b \le j \le n+1-d$,

(iii) f_{n+a+2} dominates f_j for $n+a+2 \le j \le 2n-b$,

and

		*	...	*				*	...	*	*				*	*	...	*	
		f_1	...	f_{a+1}	...	f_{n-b}	...	f_{n+1-d}	...	f_{n+a+1}	f_{n+a+2}	...	f_{2n+1-b}	...	f_{2n+1-d}	f_{2n+2-d}	...	f_{2n+1}	
*	e_1	0	...	−1	...	−1	...	−1	...	ν	ν	...	ν	...	ν	ν	...	ν	
⋮	⋮	⋮																	
*	e_{a+1}	1	...	w	...	−1	...	−1	...	−1	ν	...	ν	...	ν	ν	...	ν	($w=-1$)
	⋮	⋮																	
*	e_{n-b}	1	...	1	...	x	...	−1	...	−1	−1	...	ν	...	ν	ν	...	ν	($x=1$)
	⋮	⋮																	
*	e_{n+1-d}	1	...	1	...	1	...	0	...	−1	−1	...	−1	...	−1	ν	...	ν	
⋮	⋮	⋮																	
*	e_{n+a+1}	$-\nu$	...	1	...	1	...	1	...	0	−1	...	−1	...	−1	−1	...	−1	
	e_{n+a+2}	$-\nu$	...	$-\nu$	...	1	...	1	...	1	h	...	−1	...	−1	−1	...	−1	
	⋮	⋮																	
*	e_{2n+1-b}	$-\nu$	...	$-\nu$	...	$-\nu$	...	1	...	1	1	...	k	...	−1	−1	...	−1	
	⋮	⋮																	
	e_{2n+1-d}	$-\nu$	...	$-\nu$	...	$-\nu$	...	1	...	1	1	...	1	...	z	−1	...	−1	($z=-1$)
*	e_{2n+2-d}	$-\nu$	...	$-\nu$	...	$-\nu$	...	$-\nu$	...	1	1	...	1	...	1	0	...	−1	
⋮	⋮	⋮																	
*	e_{2n+1}	$-\nu$	...	$-\nu$	...	$-\nu$	...	$-\nu$	...	1	1	...	1	...	1	1	...	0	

Figure 13. Payoff matrix for the game of Theorem 9.4.

(iv) f_{2n+1-d} dominates f_j for $2n+1-b \le j \le 2n+1-d$.

For (i), let $a+1 \le j \le n-b$, and consider first such f_j against e_i in W_1^1, where $i \le j \le i+n$. If $i < j$ then $a_{i,j} = -1$, and if $i = j = a+1$, then $a_{i,j} = w = -1$, so all f_j in this group are equivalent against W_1^1. For e_i in W_1^2 we have $j \le i \le j+n$. If $j < i$, each $a_{i,j} = 1$, and if $j = i = n-b$ then $a_{i,j} = x = 1$, so against W_1^2 all f_j in this group are equivalent. For e_i in W_1^3 we have $i > j+n$, whence every $a_{i,j} = -\nu$. Thus, against all e_i in W_1 the f_j in this group are equivalent.

(ii) Let $n+1-b \le j \le n+1-d$, and consider first such f_j against e_i with $i \le n-b$. Then $i < j \le i+n$, so each $a_{i,j} = -1$. For e_i with $n+1-d \le i \le 2n+1-b$ we have $j \le i \le j+n$. If $j < i$ then $a_{i,j} = 1$, and if $j = i = n+1-d$ then $a_{i,j} \le 1$, so f_{n+1-d} dominates. For e_i with $i \ge 2n+2-d$ we have $i > j+n$, and every $a_{i,j} = -\nu$. Thus, against all e_i in W_1, f_{n+1-d} dominates the f_j in this group.

(iii) Let $n+a+2 \le j \le 2n-b$. For e_i in W_1^1 we have $j > i+n$, so every $a_{i,j} = \nu$. For e_i in W_1^2, $i < j \le j+n$ and every $a_{i,j} = -1$. For e_i in W_1^3, $j < i \le j+n$ and every $a_{i,j} = 1$. Thus against W_1, all f_j in this group are equivalent.

(iv) Let $2n+1-b \le j \le 2n+1-d$. For $i \le n-b$ we have all $a_{i,j} = \nu$. For $n+1-d \le i \le 2n-b$ we have $i < j \le n+i$ so that $a_{i,j} = -1$, and if $i = j = 2n+1-b$ then $a_{i,j} \ge -1$, so f_{2n+1-d} dominates in this group against all e_i in W_1 with $i \le 2n+1-b$. For the remaining e_i in W_1 we have $j < i \le j+n$, and every $a_{i,j} = 1$. Thus f_{2n+1-d} dominates the f_j in this group against all e_i in W_1, and the proof is complete. □

The next theorem deals with the single case (iiA).

THEOREM 9.5. Assume that $w = x = -1$, $y = 1$, $z = -1$, $a \le c$ and $b \ge d$. Let

$W_1^1 = \{e_i\colon 1 \le i \le a+1\} \cup \{e_{c+2}\}$,

$W_1^2 = \{e_i\colon n+1-d \le i \le n+a+1\} \cup \{e_{n+c+2}\}$,

$W_1^3 = \{e_i\colon 2n+2-d \le i \le 2n+1\}$,

$W_2^1 = \{f_i\colon 1 \le i \le a+1\}$,

$W_2^2 = \{f_i\colon n+1-d \le i \le n+a+2\}$,

$W_2^3 = \{f_i\colon 2n+1-d \le i \le 2n+1\}$,

and $W_i = W_i^1 \cup W_i^2 \cup W_i^3$, for $i = 1,2$. Then optimal strategies for the $(2a+2d+4)$ by $(2a+2d+4)$ subgame on $W_1 \times W_2$ are optimal for the full game on $\tilde{W}_1 \times \tilde{W}_2$. The reduced game is of type (9.0.1A).

PROOF. We show first that against W_2, every element of $\widetilde{W}_1 \setminus W_1$ is dominated by an element of W_1, as follows:

(i) e_{c+2} dominates all e_i with $a+2 \le i \le n-d$, and

(ii) e_{n+c+2} dominates all e_i with $n+a+2 \le i \le 2n+1-d$.

(See Figure 14 for the payoff matrix of this game.)

For (i), let $a+2 \le 1 \le n-d$. For f_j in W_2^1 we have $j < i \le j+n$ so that every $a_{i,j} = 1$, and for f_j in W_2^2, $i < j \le i+n$ and every $a_{i,j} = -1$. For f_j in W_2^3, $j > i+n$ and every $a_{i,j} = \nu$. Thus against all f_j in W_1 the e_i in this group are equivalent.

(ii) Let $n+a+2 \le i \le 2n+1-d$. For f_j in W_2^1, $i > j+n$ so that every $a_{i,j} = -\nu$. For f_j in W_2^2 we have $j \le i \le j+n$. If $j < i$ every $a_{i,j} = 1$, and if $i = j = n+a+2$ then $a_{i,j} \le 1$, so e_{n+c+2} dominates. (Note that if $a = c$ and $i = j = n+a+2$ then $a_{i,j} = y = 1$.) For f_j in W_2^3, $i \le j \le i+n$. If $i < j$ then every $a_{i,j} = -1$, and if $i = j = 2n+1-d$ then $a_{i,j} = z = -1$ also. Thus against all of W_2, e_{n+c+2} dominates the e_i in this group.

To complete the proof we show that against W_1 every element of $\widetilde{W}_2 \setminus W_2$ is dominated by one in W_2, as follows:

		f_1	…	f_{a+1}	…	f_{c+2}	…	f_{n+1-d}	…	f_{n+a+1}	f_{n+a+2}	…	f_{n+c+2}	…	f_{2n+1-d}	f_{2n+2-d}	…	f_{2n+1}	
		*	…	*				*	…	*	*				*	*	…	*	
*	e_1	0	…	−1	…	−1	…	−1	…	ν	ν	…	ν	…	ν	ν	…	ν	
⋮	⋮	⋮																	
*	e_{a+1}	1	…	w	…	−1	…	−1	…	−1	ν	…	ν	…	ν	ν	…	ν	(w=−1)
	⋮	⋮																	
*	e_{c+2}	1	…	1	…	h	…	−1	…	−1	−1	…	−1	…	ν	ν	…	ν	
	⋮	⋮																	
*	e_{n+1-d}	1	…	1	…	1	…	0	…	−1	−1	…	−1	…	−1	ν	…	ν	
⋮	⋮	⋮																	
*	e_{n+a+1}	−ν	…	1	…	1	…	1	…	0	−1	…	−1	…	−1	−1	…	−1	
	e_{n+a+2}	−ν	…	−ν	…	1	…	1	…	1	0	…	−1	…	−1	−1	…	−1	
	⋮	⋮																	
*	e_{n+c+2}	−ν	…	−ν	…	1	…	1	…	1	1	…	y	…	−1	−1	…	−1	(y=1)
	⋮	⋮																	
	e_{2n+1-d}	−ν	…	−ν	…	−ν	…	1	…	1	1	…	1	…	z	−1	…	−1	(z=−1)
*	e_{2n+2-d}	−ν	…	−ν	…	−ν	…	−ν	…	1	1	…	1	…	1	0	…	−1	
⋮	⋮	⋮																	
*	e_{2n+1}	−ν	…	−ν	…	−ν	…	−ν	…	1	1	…	1	…	1	1	…	0	

Figure 14. Payoff matrix for the game of Theorem 9.5.

(i) f_{a+1} dominates f_j for $a+1 \le j \le c+1$,

(ii) f_{n+1-d} dominates f_j for $c+2 \le j \le n+1-d$,

(iii) f_{n+a+2} dominates f_j for $n+a+2 \le j \le n+c+2$,

and

(iv) f_{2n+1-d} dominates f_j for $n+c+3 \le j \le 2n+1-d$.

For (i), let $a+1 \le j \le c+1$, and consider first such f_j against e_i with $i \le a+1$. If $i < a+1$ every $a_{i,j} = -1$, and if $i = j = a+1$ then $a_{i,j} = w = -1$, so these f_j are equivalent against this set of e_i. Next consider such f_j against e_i with $c+2 \le i \le n+a+1$. Then $j < i \le j+n$, so every $a_{i,j} = 1$. For $i \ge n+c+2$ we have $i > j+n$ and therefore every $a_{i,j} = -\nu$. Thus against all e_i in W_1 the f_j in this group are equivalent.

(ii) Let $c+2 \le j \le n+1-d$, and consider first such f_j against e_i in W_1^1, where we have $i \le j \le i+n$. If $i < j$ then every $a_{i,j} = -1$, and if $i = j = c+2$ then $a_{i,j} \ge -1$, so f_{n+1-d} dominates. Next consider such f_j against e_i in W_1^2, where we have $j \le i \le n+j$. For $j < i$, every $a_{i,j} = 1$, and if $j = i = n+1-d$ then $a_{i,j} \le 1$, so f_{n+1-d} dominates. For e_i in W_1^3 we have $i > j+n$, and every $a_{i,j} = -\nu$. Thus against all of W_1, f_{n+1-d} dominates the f_j in this group.

(iii) Let $n+a+2 \le j \le n+c+2$. For $i \le a+1$ every $a_{i,j} = \nu$, and for $c+2 \le i \le n+a+1$ we have $i < j \le i+n$, so every $a_{i,j} = -1$. For the remaining e_i in W_1 we have $j \le i \le j+n$. If $j < i$ then $a_{i,j} = 1$, and if $j = i = n+c+2$ then $a_{i,j} = y = 1$ also, so all f_j in this group are equivalent against W_1.

(iv) Let $n+c+3 \le j \le 2n+1-d$. For e_i in W_1^1 we have $j > n+1$, so every $a_{i,j} = \nu$, and for e_i in W_1^2, $i < j \le i+n$ so that every $a_{i,j} = -1$. For e_i in W_1^3, $j < i \le j+n$, and every $a_{i,j} = 1$. Thus all f_j in this group are equivalent against W_1, and the proof is complete. □

The next theorem deals with the single case (iiiD).

THEOREM 9.6. Assume that $w = x = y = -1$, $z = 1$, $a > c$ and $b < d$. Let

$W_1^1 = \{e_i\colon 1 \le i \le c+1\}$,

$W_1^2 = \{e_{n+1-d}\} \cup \{e_i\colon n+1-b \le i \le n+c+2\}$,

$W_1^3 = \{e_{2n+1-d}\} \cup \{e_i\colon 2n+1-b \le i \le 2n+1\}$,

$W_2^1 = \{f_j\colon 1 \le j \le c+2\}$,

$W_2^2 = \{f_j\colon n-b \le j \le n+c+2\}$,

$W_2^3 = \{f_j\colon 2n+1-b \le j \le 2n+1\}$,

and $W_i = W_i^1 \cup W_i^2 \cup W_i^3$ for $i = 1,2$. Then optimal strategies for the $(2b+2c+6)$ by $(2b+2c+6)$ subgame on

$W_1 \times W_2$ are optimal for the full game on $\widetilde{W}_1 \times \widetilde{W}_2$. The reduced game is of type (9.0.1D).

PROOF. We show first that against W_2, every element of $\widetilde{W}_1 \setminus W_1$ is dominated by an element of W_1, as follows:

(i) e_{n+1-d} dominates e_i for $c+2 \le i \le n-b$, and

(ii) e_{2n+1-d} dominates e_i for $n+c+3 \le i \le 2n-b$.

(See Figure 15 for the payoff matrix of the game.)

For (i), let $c+2 \le i \le n-b$, and consider first such e_i against f_j in W_2^1, where we have $j \le i \le j+n$. If $j < i$ then every $a_{i,j} = 1$, and if $j = i = c+2$ then $a_{i,j} \le 1$, so e_{n+1-d} dominates.

(ii) Let $n+c+3 \le i \le 2n-b$. For f_j in W_2^1 we have $i > j+n$ so every $a_{i,j} = -\nu$, and for f_j in W_2^2, $j < i \le j+n$ so that every $a_{i,j} = 1$. For f_j in W_2^3 we have $i < j \le i+n$ and every $a_{i,j} = -1$. Thus against W_2, all e_i in this group are equivalent.

To complete the proof we show that against W_1, each element of $\widetilde{W}_2 \setminus W_2$ is dominated by an element of W_2, as follows:

(i) f_{c+2} dominates f_j for $c+2 \le j \le n-d$,

(ii) f_{n-b} dominates f_j for $n+1-d \le j \le n-b$,

(iii) f_{n+c+2} dominates f_j for $n+c+2 \le j \le 2n+1-d$,

and

(iv) f_{2n+1-b} dominates f_j for $2n+2-d \le j \le 2n+1-b$.

		f₁	...	f_{c+1}	f_{c+2}	...	f_{n+1-d}	...	f_{n-b}	f_{n+1-b}	...	f_{n+c+2}	...	f_{2n+1-d}	...	f_{2n+1-b}	...	f_{2n+1}	
		*	...	*	*				*	*	...	*				*	...	*	
*	e_1	0	...	−1	−1	...	−1	...	−1	−1	...	ν	...	ν	...	ν	...	ν	
⋮	⋮	⋮																	
*	e_{c+1}	1	...	h	−1	...	−1	...	−1	−1	...	ν	...	ν	...	ν	...	ν	
	e_{c+2}	1	...	1	k	...	−1	...	−1	−1	...	−1	...	ν	...	ν	...	ν	
	⋮	⋮																	
*	e_{n+1-d}	1	...	1	1	...	m	...	−1	−1	...	−1	...	−1	...	ν	...	ν	
	⋮	⋮																	
	e_{n-b}	1	...	1	1	...	1	...	x	−1	...	−1	...	−1	...	ν	...	ν	(x=−1)
*	e_{n+1-b}	1	...	1	1	...	1	...	1	0	...	−1	...	−1	...	−1	...	ν	
⋮	⋮	⋮																	
*	e_{n+c+2}	$-\nu$	...	$-\nu$	1	...	1	...	1	1	...	y	...	−1	...	−1	...	−1	(y=−1)
	⋮	⋮																	
*	e_{2n+1-d}	$-\nu$	...	$-\nu$	$-\nu$	...	1	...	1	1	...	1	...	z	...	−1	...	−1	(z=1)
	⋮	⋮																	
*	e_{2n+1-b}	$-\nu$	...	$-\nu$	$-\nu$	...	$-\nu$	...	$-\nu$	1	...	1	...	1	...	0	...	−1	
⋮		⋮																	
*	e_{2n+1}	$-\nu$	...	$-\nu$	$-\nu$	...	$-\nu$	...	$-\nu$	$-\nu$	...	1	...	1	...	1	...	0	

Figure 15. Payoff matrix for the game of Theorem 9.6.

For (i), let $c+2 \le j \le n-d$. For e_i in W_1^1 we have $i < j \le i+n$, so that every $a_{i,j} = -1$. For e_i in W_1^2, $j < i \le j+n$, so every $a_{i,j} = 1$, and for e_i in W_1^3, $i > n+j$ and every $a_{i,j} = -\nu$. Thus, against W_1, all f_j in this group are equivalent.

(ii) Let $n+1-d \le j \le n-b$, and consider first such f_j against e_i with $i \le n+1-d$. If $i < j$ then every $a_{i,j} = -1$, and if $i = j = n+1-d$ then $a_{i,j} \ge -1$, so f_{n-b} dominates. Next consider such f_j against e_i with $n+1-b \le i \le 2n+1-d$. Then $j < i \le n+j$, so that every $a_{i,j} = 1$. For the remaining e_i in W_1 we have $i > n+j$, so that every $a_{i,j} = -\nu$. Thus f_{n-b} dominates the f_j in this group against all of W_1.

(iii) Let $n+c+2 \le j \le 2n+1-d$. For e_i in W_1^1, $j > n+i$ so every $a_{i,j} = \nu$. For e_i in W_1^2 we have $i \le j \le n+i$. If $i < j$ then every $a_{i,j} = -1$, and if $i = j = n+c+2$ then $a_{i,j} = y = -1$ also. For e_i in W_1^3 we have $j \le i \le j+n$. If $j < i$ then every $a_{i,j} = 1$, and if $j = i = 2n+1-d$ then $a_{i,j} = z = 1$ as well. Thus the f_j in this group are equivalent against W_1.

(iv) Let $2n+2-d \le j \le 2n+1-b$. For e_i in $W_1 \cup \{e_{n+1-b}\}$ we have $j > i+n$, so that every $a_{i,j} = \nu$. For e_i with $n+1-b \le i \le 2n+1-d$ we have $i < j \le i+n$, and every $a_{i,j} = -1$. For the remaining e_i in W_1 we have $j \le i \le j+n$. If $j < i$ then $a_{i,j} = 1$, and if $j = i = 2n+1-b$ then $a_{i,j} = 0$, so f_{2n+1-b} dominates. Thus, against all e_i in W_1, f_{2n+1-b} dominates the other f_j in this group, and the proof is complete. □

There remains only case iv, - - - -, and our next four theorems give the reduction to even order games for the subcases A ($a \le c$, $b \ge d$) and D ($a > c$, $b < d$). We begin with ivA with a + in the first part of the diagonal.

THEOREM 9.7. Assume that $w = x = y = z = -1$, $a \le c$, $b \ge d$, and that +1 occurs on the diagonal in position k, where $a+3 \le k \le n-b-2$. Let

$$W_1^1 = \{e_i\colon 1 \le i \le a+1\} \cup \{e_k\},$$

$$W_1^2 = \{e_i\colon n+1-d \le i \le n+a+1\} \cup \{e_{n+k+1}\}$$

$$W_1^3 = \{e_i\colon 2n+2-d \le i \le 2n+1\}$$

$$W_2^1 = \{f_j\colon 1 \le j \le a+1\},$$

$$W_2^2 = \{f_j\colon n+1-d \le j \le n+a+2\},$$

$$W_2^3 = \{f_j\colon 2n+1-d \le j \le 2n+1\},$$

and $W_i = W_i^1 \cup W_i^2 \cup W_i^3$ for $i = 1,2$. Then mixed strategies which are optimal for the (2a+2d+4) by (2a+2d+4) subgame on $W_1 \times W_2$ are optimal for the full game on $\tilde{W}_1 \times \tilde{W}_2$. The reduced game is of type (9.0.1A).

PROOF. The game matrix is shown in Figure 16. We show first that against W_2, every pure strategy in $\widetilde{W}_1 \setminus W_1$ is dominated by one in W_1, as follows:

(i) e_k dominates e_i for $a+2 \le i \le n-d$;

(ii) e_{n+k+1} dominates e_i for $n+a+2 \le i \le 2n+1-d$.

For (i), let $a+2 \le i \le n-d$, and consider first such strategies against f_j in W_2^1. Then $j < i \le j+n$, and thus every $a_{i,j} = 1$. For f_j in W_2^2 we have $i < j \le i+n$, and therefore every $a_{i,j} = -1$. For f_j in W_2^3, $j > n+i$ so that every $a_{i,j} = \nu$. Thus all e_i in this group are in fact equivalent against W_2.

(ii) Let $n+a+2 \le i \le 2n+1-d$, and consider first such e_i against f_j in W_2^1. Since $i > j+n$, every $a_{i,j} = -\nu$. For f_j in W_2^2 we have $j \le i \le j+n$. If $j < i$ then $a_{i,j} = 1$. If $j = i = n+a+2$, $a_{i,j} = 0$ or -1, so e_{n+k+1} dominates. For f_j in W_2^3 we have $i \le j \le i+n$. If $i < j$, every $a_{i,j} = -1$. If $i = j = 2n+1-d$, then $a_{i,j} = -1$ by hypothesis. Thus e_{n+k+1} dominates in this group against all of W_2.

To complete the proof we show that against W_1, every pure strategy in $\widetilde{W}_2 \setminus W_2$ is dominated by one in W_2, as follows:

(i) f_{a+1} dominates f_j for $a+1 \le j \le k$;

		*	...	*				*	...	*	*				*	*	...	*
		f_1	...	f_{a+1}	...	f_k	...	f_{n+1-d}	...	f_{n+a+1}	f_{n+a+2}	...	f_{n+k+1}	...	f_{2n+1-d}	f_{2n+2-d}	...	f_{2n+1}
*	e_1	0	...	−1	...	−1	...	−1	...	ν	ν	...	ν	...	ν	ν	...	ν
⋮	⋮	⋮																
*	e_{a+1}	1	...	−1	...	−1	...	−1	...	−1	ν	...	ν	...	ν	ν	...	ν
	⋮	⋮																
*	e_k	1	...	1	...	1	...	−1	...	−1	−1	...	ν	...	ν	ν	...	ν
	⋮	⋮																
*	e_{n+1-d}	1	...	1	...	1	...	0	...	−1	−1	...	−1	...	−1	ν	...	ν
⋮	⋮	⋮																
*	e_{n+a+1}	$-\nu$	...	1	...	1	...	1	...	0	−1	...	−1	...	−1	−1	...	−1
	e_{n+a+2}	$-\nu$	...	$-\nu$	...	1	...	1	...	1	h	...	−1	...	−1	−1	...	−1
	⋮	⋮																
*	e_{n+k+1}	$-\nu$	...	$-\nu$	...	$-\nu$	...	1	...	1	1	...	m	...	−1	−1	...	−1
	⋮	⋮																
	e_{2n+1-d}	$-\nu$	...	$-\nu$	...	$-\nu$	...	1	...	1	1	...	1	...	−1	−1	...	−1
*	e_{2n+2-d}	$-\nu$	...	$-\nu$	...	$-\nu$	...	$-\nu$	...	1	1	...	1	...	1	0	...	−1
⋮	⋮	⋮																
*	e_{2n+1}	$-\nu$	...	$-\nu$	...	$-\nu$	...	$-\nu$	...	1	1	...	1	...	1	1	...	0

Figure 16. Game matrix for Theorem 9.7.

(ii) f_{n+1-d} dominates f_j for $k+1 \le j \le n+1-d$;

(iii) f_{n+a+2} dominates f_j for $n+a+2 \le j \le n+k$;

(iv) f_{2n+1-d} dominates f_j for $n+k+1 \le j \le 2n+1-d$.

For (i), let $a+1 \le j \le k$, and consider first such f_j against e_i with $1 \le i \le a+1$. For $i < a+1$ we have $i < j < i+n$, so that every $a_{i,j} = -1$. If $i = j = a+1$ then $a_{i,j} = w = -1$ by hypothesis. Thus all f_j in this group are equivalent against such e_i. Next consider such f_j against e_i with $k \le i \le n+a+1$. Then $j \le i \le j+n$. If $j < i$, all $a_{i,j} = 1$, and if $j = i = k$, then $a_{i,j} = 1$ by hypothesis, so again the f_j under consideration are equivalent against these e_i. For the remaining e_i in W_1 we have $i \ge n+k+1 > j+n$ so every $a_{i,j} = -\nu$. Thus all f_j in this group are equivalent against W_1.

(ii) Let $k+1 \le j \le n+1-d$, and consider first such f_j against e_i in W_1^1. Then $i < j \le i+n$, so every $a_{i,j} = -1$. For e_i in W_1^2 we have $j \le i \le j+n$. If $j < i$ then every $a_{i,j} = 1$, and if $j = i = n+1-d$ then $a_{i,j} = 0$, so f_{n+1-d} dominates. For e_i in W_1^3 we have $i > j+n$, so every $a_{i,j} = -\nu$. Thus f_{n+1-d} dominates this set of f_j against all of W_1.

(iii) Let $n+a+2 \le j \le n+k$. For every e_i with $i \le a+1$ we have $a_{i,j} = \nu$. For e_i with $k \le i \le n+a+1$ we have $i < j \le i+n$, so every $a_{i,j} = -1$. For the remaining e_i in W_1, $j < i < j+n$ so that each $a_{i,j} = 1$. Thus these f_j are equivalent against W_1.

(iv) Let $n+k+1 \le j \le 2n+1-d$. For e_i in W_1^1 we have $j > i+n$, so every $a_{i,j} = \nu$. For e_i in W_1^2, $i \le j \le i+n$. If $i < j$, every $a_{i,j} = -1$, and if $i = j = n+k+1$, $a_{i,j} \ge -1$, so f_{2n+1-d} dominates. For e_i in W_1^3, $j < i < j+n$ and every $a_{i,j} = 1$. Thus f_{2n+1-d} dominates the f_j in this group against all of W_1, and the proof is complete. □

Subcase ivA with a + in H is handled in the next theorem.

THEOREM 9.8. Assume that $w = x = y = z = -1$, $a \le c$, $b \ge d$ and that +1 occurs on the diagonal in position $n+k$, where $c+4 \le k \le n-d-1$. Let

$W_1^1 = \{e_i\colon 1 \le i \le a+1\} \cup \{e_k\}$,

$W_1^2 = \{e_i\colon n+1-d \le i \le n+a+1\} \cup \{e_{n+k}\}$,

$W_1^3 = \{e_i\colon 2n+2-d \le i \le 2n+1\}$,

$W_2^1 = \{f_j\colon 1 \le j \le a+1\}$,

$W_2^2 = \{f_j\colon n+1-d \le j \le n+a+2\}$,

$W_2^3 = \{f_j\colon 2n+1-d \le j \le 2n+1\}$,

and $W_i = W_i^1 \cup W_i^2 \cup W_i^3$ for $i = 1,2$. Then mixed strategies which are optimal for the (2a+2d+4) by (2a+2d+4) game on $W_1 \times W_2$ are optimal for the full game on $\tilde{W}_1 \times \tilde{W}_2$. The reduced game is of type (9.0.1A).

PROOF. The game matrix is shown in Figure 17. We show first that against W_2 each pure strategy in $\tilde{W}_1 \setminus W_1$ is dominated by one in W_1, as follows:

(i) e_k dominates e_i for $a+2 \le i \le n-d$;

(ii) e_{n+k} dominates e_i for $n+a+2 \le i \le 2n+1-d$.

For (i), let $a+2 \le i \le n-d$, and consider first such e_i against f_j in W_2^1. Then $j < i < j+n$, so every $a_{i,j} = 1$. For f_j in W_2^1 we have $i < j \le i+n$, so every $a_{i,j} = -1$, and for f_j in W_2^3, $j > i+n$ so every $a_{i,j} = \nu$. Thus these e_i are equivalent against W_2.

(ii) Let $n+a+2 \le i \le 2n+1-d$, and consider first such e_i against f_j in W_2^1. Then $i > j+n$ so every $a_{i,j} = -\nu$. For f_j in W_2^2 we have $j \le i \le j+n$. If $j < i$ then $a_{i,j} = 1$, and if $j = i = n+a+2$ then $a_{i,j} \le 1$, so e_{n+k} dominates. For f_j in W_2^3 we have $i \le j \le i+n$. If $i < j$ every $a_{i,j} = -1$. If $i = j = 2n+1-d$ then $a_{i,j} = -1$ by hypothesis. Thus e_{n+k} dominates in this group against all of W_2.

To complete the proof we show that against W_1 each f_j in $\tilde{W}_2 \setminus W_2$ is dominated by one in W_2, as follows.

		*	...	*				*	...	*	*				*	*	...	*
		f_1	...	f_{a+1}	...	f_k	...	f_{n+1-d}	...	f_{n+a+1}	f_{n+a+2}	...	f_{n+k}	...	f_{2n+1-d}	f_{2n+2-d}	...	f_{2n+1}
*	e_1	0	...	−1	...	−1	...	−1	...	ν	ν	...	ν	...	ν	ν	...	ν
⋮	⋮	⋮																
*	e_{a+1}	1	...	−1	...	−1	...	−1	...	−1	ν	...	ν	...	ν	ν	...	ν
	⋮	⋮																
*	e_k	1	...	1	...	h	...	−1	...	−1	−1	...	−1	...	ν	ν	...	ν
	⋮	⋮																
*	e_{n+1-d}	1	...	1	...	1	...	0	...	−1	−1	...	−1	...	−1	ν	...	ν
⋮	⋮	⋮																
*	e_{n+a+1}	$-\nu$	...	1	...	1	...	1	...	0	−1	...	−1	...	−1	−1	...	−1
	e_{n+a+2}	$-\nu$	...	$-\nu$	...	1	...	1	...	1	m	...	−1	...	−1	−1	...	−1
	⋮	⋮																
*	e_{n+k}	$-\nu$	...	$-\nu$	...	1	...	1	...	1	1	...	1	...	−1	−1	...	−1
	⋮	⋮																
	e_{2n+1-d}	$-\nu$	...	$-\nu$	...	$-\nu$	...	1	...	1	1	...	1	...	−1	−1	...	−1
*	e_{2n+2-d}	$-\nu$	...	$-\nu$	...	$-\nu$	...	$-\nu$	...	1	1	...	1	...	1	0	...	−1
⋮	⋮	⋮																
*	e_{2n+1}	$-\nu$	...	$-\nu$	...	$-\nu$	...	$-\nu$	...	1	1	...	1	...	1	1	...	0

Figure 17. Matrix for the game of Theorem 9.8.

(i) f_{a+1} dominates f_j for $a+1 \le j \le k-1$;

(ii) f_{n+1-d} dominates f_j for $k \le j \le n+1-d$;

(iii) f_{n+a+2} dominates f_j for $n+a+2 \le j \le n+k$; and

(iv) f_{2n+1-d} dominates f_j for $n+k+1 \le j \le 2n+1-d$.

For (i), let $a+1 \le j \le k-1$, and consider first such f_j against e_i with $1 \le i \le a+1$, where we have $i \le j \le i+n$. If $i < j$ each $a_{i,j} = -1$, and if $i = j = a+1$ then $a_{i,j} = w = -1$ by hypothesis. Thus against such e_i, all f_j in this group are equivalent. Next consider such f_j against e_i with $k \le i \le n+a+1$. Then $j < i \le j+n$, so every $a_{i,j} = 1$. For the remaining e_i in W_1 we have $i > j+n$ so that every $a_{i,j} = -\nu$. Thus the f_j in this group are equivalent against all of W_1.

(ii) Let $k \le j \le n+1-d$, and consider first such f_j against e_i in W_1^1, where we have $i \le j \le i+n$. If $i < j$ every $a_{i,j} = -1$, and if $i = j = k$ then $a_{i,j} \ge -1$, so f_{n+1-d} dominates. Next consider such f_j against e_i in W_1^2, where we have $j \le i \le j+n$. If $j < i$ then each $a_{i,j} = 1$, and if $j = i = n+1-d$ then $a_{i,j} = 0$, so f_{n+1-d} dominates. Finally, for e_i in W_1^3 we have $i > j+n$ so every $a_{i,j} = -\nu$. Thus f_{n+1-d} dominates this group against all e_i in W_1.

(iii) Let $n+a+2 \le j \le n+k$, and consider first such f_j against e_i with $i \le a+1$. Then $j > i+n$ so every $a_{i,j} = \nu$. For e_i with $k \le i \le n+a+1$ we have $i < j \le n+i$ so every $a_{i,j} = -1$. For the remaining e_i in W_1 we have $j \le i \le j+n$. If $j < i$ then $a_{i,j} = 1$, and if $j = i = n+k$ then $a_{i,j} = 1$ by hypothesis, so the f_j in this group are equivalent against W_1.

(iv) Let $n+k+1 \le j \le 2n+1-d$. For e_i in W_1^1 we have $j > i+n$ so every $a_{i,j} = \nu$. For e_i in W_1^2, $i < j \le i+n$, so every $a_{i,j} = -1$, and for e_i in W_1^3 we have $j < i \le j+n$ and hence every $a_{i,j} = 1$. Thus all f_j in this group are equivalent against W_1, and the proof is complete. □

We turn now to subcase ivD, dealing first with the case of at least one + in G.

THEOREM 9.9. Assume that $w = x = y = z = -1$, $a > c$, $b < d$, and that $+1$ occurs on the diagonal in position k, where $a+3 \le k \le n-b-2$. Let

$W_1^1 = \{e_i: 1 \le i \le c+1\} \cup \{e_k\}$,

$W_1^2 = \{e_i: n+1-b \le i \le n+c+2\} \cup \{e_{n+k+1}\}$,

$W_1^3 = \{e_i: 2n+1-b \le i \le 2n+1\}$,

$W_2^1 = \{f_j: 1 \le j \le c+2\}$,

$W_2^2 = \{f_j: n-b \le j \le n+c+2\}$,

$W_2^3 = \{f_j: 2n+1-b \le j \le 2n+1\}$,

and $W_i = W_i^1 \cup W_i^2 \cup W_i^3$ for $i = 1,2$. Then mixed strategies which are optimal for the (2b+2c+6) by (2b+2c+6) game on $W_1 \times W_2$ are optimal for the full game on $\tilde{W}_1 \times \tilde{W}_2$. The reduced game is of type (9.0.1D).

PROOF. The game matrix is shown in Figure 18. We show first that against W_2 every e_i in $\tilde{W}_1 \setminus W_1$ is dominated by one in W_1, as follows.

(i) e_k dominates e_i for $c+2 \le i \le n-b$, and

(ii) e_{n+k+1} dominates e_i for $n+c+3 \le i \le 2n-b$.

For (i), let $c+2 \le i \le n-b$, and consider first such e_i against f_j in W_2^1, where we have $j \le i \le j+n$. If $j < i$ then $a_{i,j} = 1$, and if $j = i = c+2$ then $a_{i,j} \le 0$, so e_k dominates. For f_j in W_2^2 we have $i \le j \le i+n$. If $i < j$ then $a_{i,j} = -1$, and if $i = j = n-b$, $a_{i,j} = x = -1$ also. For f_j in W_2^3 we have $j > i+n$ so that every $a_{i,j} = \nu$. Thus the e_i in this group are equivalent against all f_j in W_2.

(ii) Let $n+c+3 \le i \le 2n-b$. For f_j in W_2^1 we have $i > j+n$ so every $a_{i,j} = -\nu$. For f_j in W_2^2 we have $j < i \le j+n$, so every $a_{i,j} = 1$, and for f_j in W_2^3, $i < j \le i+n$ and every $a_{i,j} = -1$. Thus the e_i in this group are likewise equivalent against all of W_2.

		*	...	*	*				*	*	...	*				*	...	*
		f_1	...	f_{c+1}	f_{c+2}	...	f_k	...	f_{n-b}	f_{n+1-b}	...	f_{n+c+2}	...	f_{n+k+1}	...	f_{2n+1-b}	...	f_{2n+1}
*	e_1	0	...	−1	−1	...	−1	...	−1	−1	...	ν	...	ν	...	ν	...	ν
⋮	⋮	⋮																
*	e_{c+1}	1	...	0	−1	...	−1	...	−1	−1	...	ν	...	ν	...	ν	...	ν
	e_{c+2}	1	...	1	h	...	−1	...	−1	−1	...	−1	...	ν	...	ν	...	ν
	⋮	⋮																
*	e_k	1	...	1	1	...	1	...	−1	−1	...	−1	...	ν	...	ν	...	ν
	⋮	⋮																
	e_{n-b}	1	...	1	1	...	1	...	−1	−1	...	−1	...	−1	...	ν	...	ν
*	e_{n+1-b}	1	...	1	1	...	1	...	1	0	...	−1	...	−1	...	−1	...	ν
⋮	⋮	⋮																
*	e_{n+c+2}	$-\nu$	...	$-\nu$	1	...	1	...	1	1	...	−1	...	−1	...	−1	...	−1
	⋮	⋮																
*	e_{n+k+1}	$-\nu$	...	$-\nu$	$-\nu$	...	$-\nu$	...	1	1	...	1	...	m	...	−1	...	−1
	⋮	⋮																
*	e_{2n+1-b}	$-\nu$	...	$-\nu$	$-\nu$	...	$-\nu$	...	$-\nu$	1	...	1	...	1	...	0	...	−1
⋮	⋮	⋮																
*	e_{2n+1}	$-\nu$	...	$-\nu$	$-\nu$	...	$-\nu$	...	$-\nu$	$-\nu$	...	1	...	1	...	1	...	0

Figure 18. Game matrix for Theorem 9.9.

To complete the proof we show that against W_1, every f_j in $\widetilde{W}_2 \setminus W_2$ is dominated by one in W_2, as follows.

(i) f_{c+2} dominates f_j for $c+2 \le j \le k$;

(ii) f_{n-b} dominates f_j for $k+1 \le j \le n-b$;

(iii) f_{n+c+2} dominates f_j for $n+c+2 \le j \le n+k$; and

(iv) f_{2n+1-b} dominates f_j for $n+k+1 \le j \le 2n+1-b$.

For (i), let $c+2 \le j \le k$. If $i \le c+1$ then $i < j < i+n$ and every $a_{i,j} = -1$. For $k \le i \le n+c+2$ we have $j \le i \le j+n$. If $j < i$ then every $a_{i,j} = 1$, and if $j = i = k$ then $a_{i,j} = 1$ by hypothesis. For the remaining e_i in W_1 we have $i > j+n$ so that every $a_{i,j} = -\nu$. Thus the f_j in this group are equivalent against W_1.

(ii) Let $k+1 \le j \le n-b$, and consider first such f_j against e_i in W_1^1. Then $i < j \le i+n$, so every $a_{i,j} = -1$. For e_i in W_1^2 we have $j < i \le j+n$ so that every $a_{i,j} = 1$, and for e_i in W_1^3, $i > j+n$ so every $a_{i,j} = -\nu$. Thus the f_j in this group are equivalent against W_1.

(iii) Let $n+c+2 \le j \le n+k$. For $1 \le i \le c+1$ every $a_{i,j} = \nu$, since $j > i+n$. For $k \le i \le n+c+2$ we have $i \le j \le i+n$. If $i < j$ then every $a_{i,j}$ is -1, and if $i = j = n+c+2$ then $a_{i,j} = y = -1$ also. For the remaining e_i in W_1 we have $j < i < j+n$ so that every $a_{i,j} = 1$. Thus the f_j in this group are equivalent against W_1.

(iv) Let $n+k+1 \le j \le 2n+1-b$. For e_i in W_1^1 we have $j > i+n$ and hence every $a_{i,j} = \nu$. For e_i in W_1^2 we have $i \le j \le i+n$. Each $a_{i,j}$ with $i < j$ is -1, and if $i = j = n+k+1$ then $a_{i,j} \ge -1$, so f_{2n+1-b} dominates. For e_i in W_1^3 we have $j \le i \le j+n$. If $j < i$ then each $a_{i,j} = 1$, and if $i = j = 2n+1-b$ then $a_{i,j} = 0$ (since $b < d$). Thus f_{2n+1-b} dominates the f_j in this group against all of W_1, and the proof is complete. □

Our final theorem covers subcase ivD with at least one + in H.

THEOREM 9.10. Assume that $w = x = y = z = -1$, $a > c$, $b < d$, and that for some k with $c+4 \le k \le n-d-1$, +1 occurs on the diagonal in position n+k. Let

$W_1^1 = \{e_i\colon 1 \le i \le c+1\} \cup \{e_k\}$,

$W_1^2 = \{e_i\colon n+1-b \le i \le n+c+2\} \cup \{e_{n+k}\}$,

$W_1^3 = \{e_i\colon 2n+1-b \le i \le 2n+1\}$,

$W_2^1 = \{f_j\colon 1 \le j \le c+2\}$,

$W_2^2 = \{f_j\colon n-b \le j \le n+c+2\}$,

$W_2^3 = \{f_j\colon 2n+1-b \le j \le 2n+1\}$,

and $W_i = W_i^1 \cup W_i^2 \cup W_i^3$ for $i = 1,2$. Then mixed strategies which are optimal for the (2b+2c+6) by (2b+2c+6) game on $W_1 \times W_2$ are optimal for the full game on $\widetilde{W}_1 \times \widetilde{W}_2$. The reduced game is of type (9.0.1D).

PROOF. The game matrix is shown in Figure 19. We show first that against W_2 every element of $\tilde{W}_1 \setminus W_1$ is dominated by one in W_1, as follows.

(i) e_k dominates e_i for $c+2 \le i \le n-b$, and

(ii) e_{n+k} dominates e_i for $n+c+3 \le i \le 2n-b$.

For (i), let $c+2 \le i \le n-b$, and consider first such e_i against f_j in W_2^1, where we have $j \le i \le j+n$. If $j < i$ then $a_{i,j} = 1$, and if $i = j = c+2$ then $a_{i,j} \le 0$, so e_k dominates. For f_j in W_2^2 we have $i \le j \le i+n$. If $i < j$ then $a_{i,j} = -1$, and if $i = j = n-b$ then $a_{i,j} = x = -1$ also. For f_j in W_2^3 we have $j > i+n$ so that every $a_{i,j} = \nu$. Thus e_k dominates this group of e_i against all of W_2.

(ii) Let $n+c+3 \le i \le 2n-b$. For f_j in W_2^1, $i > j+n$ so every $a_{i,j} = -\nu$. For f_j in W_2^2, $j < i \le j+n$, so every $a_{i,j} = 1$, and for f_j in W_2^3 we have $i < j \le i+n$ and hence every $a_{i,j} = -1$. Thus the e_i in this group are equivalent against W_2.

To complete the proof we show that against W_1 every f_j in $\tilde{W}_2 \setminus W_2$ is dominated by one in W_2, as follows:

(i) f_{c+2} dominates f_j for $c+2 \le j \le k-1$;

(ii) f_{n-b} dominates f_j for $k \le j \le n-b$;

		*	...	*	*				*	*	...	*				*	...	*
		f_1	...	f_{c+1}	f_{c+2}	...	f_k	...	f_{n-b}	f_{n+1-b}	...	f_{n+c+2}	...	f_{n+k}	...	f_{2n+1-b}	...	f_{2n+1}
*	e_1	0	...	−1	−1	...	−1	...	−1	−1	...	ν	...	ν	...	ν	...	ν
⋮	⋮	⋮																
*	e_{c+1}	1	...	0	−1	...	−1	...	−1	−1	...	ν	...	ν	...	ν	...	ν
	e_{c+2}	1	...	1	h	...	−1	...	−1	−1	...	−1	...	ν	...	ν	...	ν
	⋮	⋮																
*	e_k	1	...	1	1	...	m	...	−1	−1	...	−1	...	−1	...	ν	...	ν
	⋮	⋮																
	e_{n-b}	1	...	1	1	...	1	...	−1	−1	...	−1	...	−1	...	ν	...	ν
*	e_{n+1-b}	1	...	1	1	...	1	...	1	0	...	−1	...	−1	...	−1	...	ν
⋮	⋮	⋮																
*	e_{n+c+2}	$-\nu$	...	$-\nu$	1	...	1	...	1	1	...	−1	...	−1	...	−1	...	−1
	⋮	⋮																
*	e_{n+k}	$-\nu$	...	$-\nu$	$-\nu$	...	1	...	1	1	...	1	...	1	...	−1	...	−1
	⋮	⋮																
*	e_{2n+1-b}	$-\nu$	...	$-\nu$	$-\nu$	...	$-\nu$	...	$-\nu$	1	...	1	...	1	...	0	...	−1
⋮	⋮	⋮																
*	e_{2n+1}	$-\nu$	...	$-\nu$	$-\nu$	...	$-\nu$	...	$-\nu$	$-\nu$	...	1	...	1	...	1	...	0

Figure 19. Matrix for the game of Theorem 9.10.

(iii) f_{n+c+2} dominates f_j for $n+c+2 \le j \le n+k$; and

(iv) f_{2n+1-b} dominates f_j for $n+k+1 \le j \le 2n+1-b$.

For (i), let $c+2 \le j \le k-1$, and consider first such f_j against e_i with $1 \le i \le c+1$. Then $i < j \le i+n$ so every $a_{i,j} = -1$. Against e_i with $k \le i \le n+c+2$ these f_j are again equivalent, since $j < i \le j+n$, so that every $a_{i,j} = 1$. For the remaining e_i in W_1 we have $i > j+n$, so every $a_{i,j} = -\nu$. Thus, against all of W_1 the f_j in this group are equivalent.

(ii) Let $k \le j \le n-b$, and consider first such f_j against e_i in W_1^1, where we have $i \le j \le i+n$. If $i < j$, every $a_{i,j} = -1$, and if $i = j = k$ then $a_{i,j} \ge -1$, so f_{n-b} dominates. For e_i in W_1^2 we have $j < i \le j+n$, so every $a_{i,j} = 1$, and for e_i in W_1^3, $i > j+n$, so that every $a_{i,j} = -\nu$. Thus f_{n-b} dominates the f_j in this group against all of W_1.

(iii) Let $n+c+2 \le j \le n+k$, and consider first such f_j against e_i with $1 \le i \le c+1$. Then $j > i+n$, so every $a_{i,j} = \nu$. Next consider such f_j against e_i with $k \le i \le n+c+2$, in which case we have $i \le j \le i+n$. If $i < j$ then $a_{i,j} = -1$, and if $i = j = n+c+2$ then $a_{i,j} = y = -1$ also. For the remaining e_i in W_1, we have $j \le i \le j+n$. If $j < i$ then every $a_{i,j} = 1$, and if $j = i =$

n+k, then $a_{i,j} = 1$ by hypothesis. Thus all f_j in this group are equivalent against W_1.

(iv) Let $n+k+1 \le j \le 2n+1-b$. For e_i in W_1^1 we have $j > i+n$, so every $a_{i,j} = \nu$. For e_i in W_1^2, $i < j \le i+n$, so every $a_{i,j} = -1$. For e_i in W_1^3 we have $j \le i \le j+n$. If $j < i$ then every $a_{i,j} = 1$, and if $j = i = 2n+1-b$ then $a_{i,j} = 0$, so f_{2n+1-b} dominates. Thus f_{2n+1-b} dominates the f_j in this group against all of W_1, and the proof is complete. □

10. Games with ±1 as central diagonal element.

When the central diagonal element is ±1, the facts are considerably simpler. It again appears to be the case that unless both +1 and -1 occur on the diagonal, the game is irreducible. We shall show that when both do occur, the game always reduces to the 2 by 2 game $\begin{bmatrix} -1 & \nu \\ 1 & -1 \end{bmatrix}$ or $\begin{bmatrix} 1 & -1 \\ -\nu & 1 \end{bmatrix}$ according as the central diagonal element is +1 or -1. Let us denote the diagonal elements $(x_1, x_2, \ldots, x_{2n+1})$.

THEOREM 10.1. Assume that $x_{n+1} = +1$ and that for some $k < n$, $x_k = -1$. Let $W_1 = \{e_1, e_{n+1}\}$ and $W_2 = \{f_k, f_{n+k+1}\}$. Then optimal strategies for the subgame on $W_1 \times W_2$ are optimal for the full game on $\hat{W}_1 \times \hat{W}_2$. These optimal strategies are $P = (2,\nu+1)/(\nu+3)$, $Q = (\nu+1,2)/(\nu+3)$, and the game value is $(\nu-1)/(\nu+3)$.

PROOF. It is easy to see that the matrix for the game on $W_1 \times W_2$ is $\begin{bmatrix} -1 & \nu \\ 1 & -1 \end{bmatrix}$, and that the optimal strategies and game value for this game are as asserted. We show now that these strategies are optimal for the full game by showing that $E(P,f_j) \geq V$ for every f_j in $\hat{W}_2$ and $E(e_i,Q) \leq V$ for every e_i in $\hat{W}_1$, where $V = (\nu-1)/\nu+3)$. See Figure 20 for the matrix of the full game.

For $j \le n+1$ we have $a_{1,j} \ge -1$ and $a_{n+1,j} = 1$, so $E(P,f_j) = [2a_{1,j} + (\nu+1)a_{n+1,j}]/(\nu+3) \ge [-2 + (\nu+1)]/(\nu+3) = V$. For $j > n+1$, $a_{1,j} = \nu$ and $a_{n+1,j} = -1$, so $E(P,f_j) = [2\nu - (\nu+1)]/(\nu+3) = V$.

Now consider $E(e_i,Q)$ for $i \le k$. If $i < k$ then $a_{i,k} = -1$, and $a_{k,k} = -1$ by hypothesis. For all $i \le k$, $a_{i,n+k+1} = \nu$, so $E(e_i,Q) = [(\nu+1)\ a_{i,k} + 2a_{i,n+k+1}]/(\nu+3) =$

				$(\nu+1)$						(2)		
		f_1	...	f_k	...	f_n	f_{n+1}	f_{n+2}	...	f_{n+k+1}	...	f_{2n+1}
(2)	e_1	x_1	...	-1	...	-1	-1	ν	...	ν	...	ν
	⋮	⋮										
	e_k	1	...	-1	...	-1	-1	-1	...	ν	...	ν
	⋮	⋮										
	e_n	1	...	1	...	x_n	-1	-1	...	-1	...	ν
$(\nu+1)$	e_{n+1}	1	...	1	...	1	1	-1	...	-1	...	-1
	e_{n+2}	$-\nu$	...	1	...	1	1	x_{n+2}	...	-1	...	-1
	⋮	⋮										
	e_{n+k+1}	$-\nu$	...	$-\nu$	...	1	1	1	...	x_{n+k+1}	...	-1
	⋮	⋮										
	e_{2n+1}	$-\nu$	...	$-\nu$	...	$-\nu$	1	1	...	1	...	x_{2n+1}

Figure 20. Game matrix for Theorem 10.1

$[-(\nu+1) + 2\nu]/(\nu+3) = V$. Next consider $k < i \le n+k$. Then $a_{i,k} = 1$ and $a_{i,n+k+1} = -1$, so $E(e_i,Q) =$

$[(\nu+1)-2]/(\nu+3) = V$. Finally, for $i > n+k$ we have $a_{i,k} = -\nu$ and $a_{i,n+k+1} \leq 1$. Thus $E(e_i,Q) \leq [-\nu(\nu+1) + 2]/(\nu+3) = -(\nu+2)(\nu-1)/(\nu+3) < 0 \leq V$, and the proof is complete. □

If $x_{n+1} = -1$ and for some $k < n$, $x_k = +1$, then we have the game of Theorem 10.1 with the roles of the players reversed. We now deal with the case where $x_{n+1} = -1$ and $+1$ occurs on the right half of the diagonal.

THEOREM 10.2 Assume that $x_{n+1} = -1$ and that $x_{n+k} = +1$ for some k, $3 \leq k \leq n+1$. Let $W_1 = \{e_k, e_{n+k}\}$ and $W_2 = \{f_1, f_{n+1}\}$. Then optimal strategies for the subgame on $W_1 \times W_2$ are optimal for the full game on $\widetilde{W}_1 \times \widetilde{W}_2$. These optimal strategies are $P = (\nu+1,2)/(\nu+3)$, $Q = (2,\nu+1)/(\nu+3)$, and the game value is $(-\nu+1)/(\nu+3)$.

PROOF. Observe that the matrix of the game on $W_1 \times W_2$ is $\begin{bmatrix} 1 & -1 \\ -\nu & 1 \end{bmatrix}$. One checks readily that the optimal strategies and value for this game are as asserted. We show that they are optimal for the full game by showing that $E(P,f_j) \geq V$ for every f_j in $\widetilde{W}_2$ and $E(e_i,Q) \leq V$ for every e_i in $\widetilde{W}_1$, where $V = (-\nu+1)/(\nu+3)$. The matrix of the game is shown in Figure 21.

For $j < k$ each $a_{k,j} = 1$ and $a_{n+k,j} = -\nu$, so $E(P,f_j) = [(\nu+1)a_{k,j} + 2a_{n+k,j}]/(\nu+3) = [(\nu+1) - 2\nu]/(\nu+3) = V$. For $k \le j \le n+k$, $a_{k,j} \ge -1$ and $a_{n+k,j} = 1$, so $E(P,f_j) \ge [-(\nu+1) + 2]/(\nu+3) = V$. For $j > n+k$, $a_{k,j} = \nu$ and $a_{n+k,j} = -1$. Then $E(P,f_j) = [\nu(\nu+1) - 2]/(\nu+3) = (\nu+2)(\nu-1)/(\nu+3) > 0 \ge V$, so we have $E(P,f_j) \ge V$ for every f_j in $\widetilde{W}_2$.

		(2)					(ν+1)					
		f_1	…	f_k	…	f_n	f_{n+1}	f_{n+2}	…	f_{n+k}	…	f_{2n+1}
	e_1	x_1	…	-1	…	-1	-1	ν	…	ν	…	ν
	⋮	⋮										
(ν+1)	e_k	1	…	x_k	…	-1	-1	-1	…	-1	…	ν
	⋮	⋮										
	e_n	1	…	1	…	x_n	-1	-1	…	-1	…	ν
	e_{n+1}	1	…	1	…	1	-1	-1	…	-1	…	-1
	e_{n+2}	$-\nu$	…	1	…	1	1	x_{n+2}	…	-1	…	-1
	⋮	⋮										
(2)	e_{n+k}	$-\nu$	…	1	…	1	1	1	…	1	…	-1
	⋮	⋮										
	e_{2n+1}	$-\nu$	…	$-\nu$	…	$-\nu$	1	1	…	1	…	x_{2n+1}

Figure 21. Matrix for the game of Theorem 10.2.

Now consider $E(e_i,Q)$. For $i \leq n+1$, every $a_{i,1} \leq 1$ and $a_{i,n+1} = -1$. Thus $E(e_i,Q) = [2a_{i,1} + (\nu+1)a_{i,n+1}]/(\nu+3) \leq [2 - (\nu+1)]/(\nu+3) = V$. For $i > n+1$, $a_{i,1} = -\nu$ and $a_{i,n+1} = 1$, so $E(e_i,Q) = [-2\nu + (\nu+1)]/(\nu+3) = V$. Thus $E(e_i,Q) \leq V$ for every e_i in $\widetilde{W}_1$, and the proof is complete. □

11. Further reduction to 2 by 2 when $\nu = 1$.

We show now how all of the reduced games in Sections 8 and 9 reduce further, if $\nu = 1$, to 2 by 2 games with matrix

$$A_0 = \begin{bmatrix} 1 & -1 \\ -1 & 1 \end{bmatrix}. \tag{11.0.1}$$

This is the matrix A' of Section 3, with $\nu = 1$.

The optimal strategies and game value are

$$P = Q = (.5, .5) \ , \ V = 0. \tag{11.0.2}$$

Recall that all games in Section 8 reduce to balanced games with one of the four diagonals (8.0.5A) to (8.0.5D). Our first theorem below shows how all of these reduce to 2 by 2 when $\nu = 1$.

THEOREM 11.1. Let $\tilde{W}_1 = \{e_1, e_2, \ldots, e_{2n+1}\}$ and $\tilde{W}_2 = \{f_1, f_2, \ldots, f_{2n+1}\}$ be the strategy sets in a balanced Silverman game with one of the diagonals (8.0.5A) to (8.0.5D). Let

$W_1 = \{e_{a+2}, e_{n+a+2}\}$, $W_2 = \{f_{a+1}, f_{n+a+2}\}$ in case (A) or (C);

$W_1 = \{e_{c+2}, e_{n+c+2}\}$, $W_2 = \{f_{c+2}, f_{n+c+3}\}$ in case (B) or (D).

Then for $\nu = 1$ the game may be reduced to the 2 by 2 game on $W_1 \times W_2$, having the matrix and solution given in (11.0.1) and (11.0.2).

PROOF. For cases (A) and (C) the payoff matrix is shown in Figure 22, where the entry u is 0 in case

(A) and is -1 in case (C). One sees that against W_2, each of the strategies e_i, $a+2 \le i \le n+a+1$, is equivalent to e_{a+2}, and each e_i with $i < a+2$ or $i > n+a+1$ is equivalent to e_{n+a+2} if $\nu = 1$. Against W_1, each of the strategies f_j, $a+2 \le j \le n+a+2$, is dominated by f_{n+a+2}, and each of the remaining f_j is equivalent to f_{a+1} when $\nu = 1$. Thus, optimal strategies for the game on $W_1 \times W_2$ are optimal for the full game on $\tilde{W}_1 \times \tilde{W}_2$.

	f_1	...	* f_{a+1}	f_{a+2}	...	f_{n+a+1}	* f_{n+a+2}	...	f_{2n+1}
e_1	0	...	-1	-1	...	ν	ν	...	ν
⋮	⋮								
e_{a+1}	1	...	-1	-1	...	-1	ν	...	ν
* e_{a+2}	1	...	1	u	...	-1	-1	...	ν
⋮	⋮								
e_{n+a+1}	$-\nu$	...	1	1	...	0	-1	...	-1
* e_{n+a+2}	$-\nu$	...	$-\nu$	1	...	1	1	...	-1
⋮	⋮								
e_{2n+1}	$-\nu$	...	$-\nu$	$-\nu$	...	1	1	...	0

$$u = \begin{cases} 0 & \text{in (A)} \\ -1 & \text{in (C)} \end{cases}$$

Figure 22. Payoff matrix for game of Theorem 11.1 (A) and (C).

The payoff matrix for cases (B) and (D) is shown in Figure 23, where the entry u is 1 in case (B) and is 0 in case (D). One sees that against W_2 the strategies e_i with $c+3 \le i \le n+c+2$ are all

	f_1	...	f_{c+1}	* f_{c+2}	f_{c+3}	...	f_{n+c+2}	* f_{n+c+3}	...	f_{2n+1}
e_1	0	...	−1	−1	−1	...	ν	ν	...	ν
⋮	⋮									
e_{c+1}	1	...	0	−1	−1	...	ν	ν	...	ν
* e_{c+2}	1	...	1	−1	−1	...	−1	ν	...	ν
e_{c+3}	1	...	1	1	0	...	−1	−1	...	ν
⋮	⋮									
* e_{n+c+2}	$-\nu$	...	$-\nu$	1	1	...	1	−1	...	−1
e_{n+c+3}	$-\nu$	...	$-\nu$	$-\nu$	1	...	1	u	...	−1
⋮	⋮									
e_{2n+1}	$-\nu$	...	$-\nu$	$-\nu$	$-\nu$	...	1	1	...	0

$$u = \begin{cases} 1 \text{ in } (B) \\ 0 \text{ in } (D) \end{cases}$$

Figure 23. Payoff matrix for game of Theorem 11.1 (B) and (D).

equivalent, and the remaining e_i are dominated by e_{c+2} if $\nu = 1$. Against W_1 the strategies f_j with $c+2 \le j \le n+c+2$ are equivalent to f_{c+2}, and when $\nu = 1$ the other f_j are equivalent to f_{n+c+3}. Thus, optimal strategies for the game on $W_1 \times W_2$ are optimal for the full game. It is easy to check that this 2 by 2 subgame has the matrix and solution asserted. □

All games in Section 9 reduce to even order games having matrix format as shown in Figure 9, and having one of the four main diagonal and subdiagonal configurations (9.0.1A) to (9.0.1D). We drop the asterisks now from n and s. The payoff function outside the main diagonal and first subdiagonal is given by

$$(11.1.1) \qquad A(e_i, f_j) = \begin{cases} \nu \text{ if } j \geq i+n \\ -1 \text{ if } i < j < i+n \\ 1 \text{ if } j+1 < i \leq j+n \\ -\nu \text{ if } i > j+n \end{cases} .$$

For $j \leq i \leq j+1$, $A(e_i, f_j)$ is specified in each case by the given main diagonal and subdiagonal.

THEOREM 11.2. Let $\tilde{W}_1 = \{e_1, e_2, \ldots, e_{2n}\}$ and $\tilde{W}_2 = \{f_1, f_2, \ldots, f_{2n}\}$ be strategy sets with payoff function A given by (11.1.1) and one of the diagonal-subdiagonal configurations (9.0.1A) to (9.0.1D). Let

$W_1 = \{e_{a+2}, e_{n+a+2}\}$, $W_2 = \{f_{a+1}, f_{n+a+1}\}$ in case (A) or (C);

$W_1 = \{e_{c+2}, e_{n+c+2}\}$, $W_2 = \{f_{c+2}, f_{n+c+2}\}$ in case (B) or (D).

Then for $\nu = 1$ the game may be reduced to the 2 by 2 game on $W_1 \times W_2$, having the matrix and solution given in (11.0.1) and (11.0.2).

PROOF. For cases (A) and (C) the payoff matrix is shown in Figure 24, where the element u is -1 in case (A) and 0 in case (C). The zeros on the subdiagonal are irrelevant to the proof. The relevant subdiagonal entries are $A(e_{a+2}, f_{a+1}) = 1$ and $A(e_{n+a+2}, f_{n+a+1}) = 1$. Against W_2, the strategies e_i with $a+2 \le i \le n+a+1$ are all equivalent to e_{a+2}, and with $\nu = 1$ each of the remaining e_i is equivalent to e_{n+a+2}. Against W_1, each f_j with $a+2 \le j \le n+a+1$ is equivalent to f_{n+a+1}, and with $\nu = 1$ the remaining strategies f_j

			*			*			
	f_1	...	f_{a+1}	f_{a+2}	...	f_{n+a+1}	f_{n+a+2}	...	f_{2n}
e_1	0	...	-1	-1	...	ν	ν	...	ν
⋮	⋮								
e_{a+1}	1	...	-1	-1	...	ν	ν	...	ν
* e_{a+2}	1	...	1	-1	...	-1	ν	...	ν
⋮	⋮								
e_{n+a+1}	$-\nu$	...	1	1	...	-1	-1	...	-1
* e_{n+a+2}	$-\nu$	...	$-\nu$	1	...	1	u	...	-1
⋮	⋮								
e_{2n}	$-\nu$	...	$-\nu$	$-\nu$	...	1	1	...	0

$$u = \begin{cases} -1 & \text{in (A)} \\ 0 & \text{in (C)} \end{cases}$$

Figure 24. Payoff matrix for game of Theorem 11.2 (A) and (C).

are dominated by f_{a+1}. Thus, optimal strategies for the game on $W_1 \times W_2$ are optimal for the full game.

For cases (B) and (D) the payoff matrix is shown in Figure 25. One sees that against W_2, the strategies e_i with $c+3 \le i \le n+c+2$ are dominated by e_{n+c+2}, and with $\nu = 1$ each of the remaining e_1 is equivalent to e_{c+2}. Against W_1, each f_j with $c+2 \le j \le n+c+1$ is

	f_1	...	f_{c+1}	* f_{c+2}	f_{c+3}	...	f_{n+c+1}	* f_{n+c+2}	f_{n+c+3}	...	f_{2n}
e_1	0	...	-1	-1	-1	...	ν	ν	ν	...	ν
⋮	⋮										
e_{c+1}	1	...	0	-1	-1	...	ν	ν	ν	...	ν
* e_{c+2}	1	...	1	-1	-1	...	-1	ν	ν	...	ν
e_{c+3}	1	...	1	u	-1	...	-1	-1	ν	...	ν
⋮	⋮										
e_{n+c+1}	$-\nu$	...	1	1	1	...	-1	-1	-1	...	-1
* e_{n+c+2}	$-\nu$	...	$-\nu$	1	1	...	1	-1	-1	...	-1
e_{n+c+3}	$-\nu$	...	$-\nu$	$-\nu$	1	...	1	1	0	...	-1
⋮	⋮										
e_{2n}	$-\nu$	...	$-\nu$	$-\nu$	$-\nu$	...	1	1	1	...	0

$$u = \begin{cases} 0 \text{ in (B)} \\ 1 \text{ in (D)} \end{cases}$$

Figure 25. Payoff matrix for game of Theorem 11.2 (B) and (D).

equivalent to f_{c+2}, and with $\nu = 1$ each of the remaining f_j is equivalent to e_{n+c+2}. Thus optimal strategies for the game on $W_1 \times W_2$ are optimal for the full game.

It is easy to see that in all cases the reduced game is as asserted in the theorem. □

12. Explicit solutions for certain classes.

In the papers [2] on symmetric games and [7] on disjoint games, explicit optimal strategies and game values are obtained for all games. The fact that the diagonal consists entirely of zeros in the symmetric case and entirely of ones in the disjoint case has the effect that the components in the optimal strategy vectors may be described by simple recursions. For nonconstant diagonals these relations among the components are less regular, but in a few cases where the diagonal is nearly constant one can still obtain relatively nice explicit formulas. We shall do so here for diagonals which are constant except for the middle element, or constant except for the last element.

The notation $\alpha = 2/(\nu+1)$ used in [7] will be useful again here. We first treat the games with diagonal $(-1 \ldots -1\ 0\ -1 \ldots -1)$, the zero being the central diagonal element.

THEOREM 12.1. In the balanced $2n+1$ by $2n+1$ Silverman game with central diagonal element 0 and all other diagonal elements equal to -1, the game value is

$$V = \left(\sum_{j=2}^{n} \alpha^{2j-1} - \sum_{j=1}^{n} \alpha^{2j} \right)/D, \text{ where } D = 1 + \alpha + \sum_{j=0}^{2n} \alpha^{j} ,$$

and optimal mixed strategies for the row and column players, respectively, are P/D and Q/D, where

$$P = (\alpha^{2n}+\alpha, \alpha^{2n-2}, \alpha^{2n-4}, \ldots, \alpha^{2}, 2, \alpha^{2n-1}, \alpha^{2n-3}, \ldots, \alpha);$$

$$Q = (\alpha, \alpha^{3}, \ldots, \alpha^{2n-1}, 2, \alpha^{2}, \alpha^{4}, \ldots, \alpha^{2n-2}, \alpha^{2n}+\alpha).$$

PROOF. We show that $PA = DV(1,1,\ldots,1)$, $AQ^t = DV(1,1,\ldots,1)^t$, where A is the payoff matrix, and the theorem follows.

Let C_j denote the j-th column of A, and P_i the i-th component of P. Then

$$PC_{n+1} = -\sum_{i=1}^{n+1} p_i + \sum_{i=n+2}^{2n+1} p_i = -\sum_{i=1}^{n} \alpha^{2j} + \sum_{i=2}^{n} \alpha^{2j-1} = DV.$$

Also, $P(C_{n+1}-C_n) = - p_{n+1} + (\nu+1)p_{2n+1} = -2 + (\nu+1)\alpha = 0.$

For j = 1 to n - 1,

$P(C_{j+1}-C_j) = -2p_{j+1} + (\nu+1)p_{j+n+1} = -2\alpha^{2n-2j} + (\nu+1)\alpha^{2n-2j+1} = 0$, so we have $PC_j = DV$ for $1 \leq j \leq n+1$.

Next we have

$$\begin{aligned} P(C_{n+2}-C_{n+1}) &= (\nu+1)p_1 - p_{n+1} - 2p_{n+2} \\ &= (\nu+1)(\alpha^{2n}+\alpha) - 2 - 2\alpha^{2n-1} \\ &= 0 \text{ since } (\nu+1)\alpha = 2. \end{aligned}$$

For j = 2 to n we have

$$P(C_{n+j+1}-C_{n+j}) = (\nu+1)p_j - 2p_{n+j+1}$$

$$= (\nu+1)\alpha^{2n-2j+2} - 2\alpha^{2n-2j+1} = 0,$$

and thus $PC_j = DV$ for $1 \le j \le 2n+1$.

We turn now to AQ^t, and denote by R_i the i-th row of A; q_i is the i-th component of Q. Clearly $R_{n+1}Q^t = PC_{n+1} = DV$. Also,

$$(R_{n+1}-R_n)Q^t = 2q_n + q_{n+1} - (\nu+1)q_{2n+1}$$

$$= 2\alpha^{2n-1} + 2 - (\nu+1)(\alpha^{2n}+\alpha) = 0.$$

For $1 \le j \le n-1$,

$$(R_{j+1}-R_j)Q^t = 2q_j - (\nu+1)q_{j+n+1}$$

$$= 2\alpha^{2j-1} - (\nu+1)\alpha^{2j} = 0.$$

Note next that

$$(R_{n+2}-R_{n+1})Q^t = -(\nu+1)q_1 + q_{n+1}$$

$$= -(\nu+1)\alpha + 2 = 0,$$

and for $2 \le j \le n$,

$$(R_{n+j+1}-R_{n+j})Q^t = -(\nu+1)q_j + 2q_{n+j}$$

$$= -(\nu+1)\alpha^{2j-1} + 2\alpha^{2j-2} = 0.$$

Thus $R_iQ^t = DV$ for all i, $1 \le i \le 2n+1$, and the proof is complete. □

The next theorem deals with games having diagonal (-1 -1 ... -1 0).

THEOREM 12.2. In the balanced 2n+1 by 2n+1 Silverman game with last diagonal element equal to 0 and all other diagonal elements equal to -1, the game value is

$$V = \left(2\alpha - 2 + \sum_{j=2}^{n} \alpha^{2j-1} - \sum_{j=1}^{n} \alpha^{2j}\right)/D,$$

$$\text{where } D = 1 + \alpha + \sum_{j=0}^{2n} \alpha^{j},$$

and optimal strategies for the row and column players, respectively, are P/D and Q/D, where

$$P = (\alpha^{2n}, \alpha^{2n-2}, \ldots, \alpha^{2}, 2, \alpha^{2n-1}, \alpha^{2n-3}, \ldots, \alpha^{3}, 2\alpha);$$

$$Q = (\alpha\beta, \alpha^{3}\beta, \ldots, \alpha^{2n-3}\beta, 2\alpha^{2n-1}, \beta, \alpha^{2}\beta, \ldots, \alpha^{2n-2}\beta, 2\alpha^{2n}),$$

where $\beta = 2-\alpha^2$.

PROOF. Again we shall show that each component of PA and each component of AQ^t is DV. We again denote the j-th column of A by C_j, and the i-th row by R_i. We note first that

$$PC_{n+1} = -\sum_{i=1}^{n+1} p_i + \sum_{i=n+2}^{2n+1} p_i$$

$$= -\sum_{j=1}^{n} \alpha^{2j} - 2 + \sum_{j=2}^{n} \alpha^{2j-1} + 2\alpha = DV.$$

For $1 \le j \le n$, $P(C_{j+1}-C_j) = -2p_{j+1} + (\nu+1)p_{n+j+1}$. If $j = n$, this amounts to $-4 + 2(\nu+1)\alpha = 0$, and if $j < n$, it is $-2\alpha^{2n-2j} + (\nu+1)\alpha^{2n-2j+1} = 0$. For $1 \le j \le n-1$,

$$P(C_{n+j+1}-C_{n+j}) = (\nu+1)p_j - 2p_{n+j+1}$$

$$= (\nu+1)\alpha^{2n-2j+2} - 2\alpha^{2n-2j+1} = 0,$$

and $P(C_{2n+1}-C_{2n}) = (\nu+1)p_n - p_{2n+1}$

$$= (\nu+1)\alpha^2 - 2\alpha = 0.$$

Thus we have $PC_j = DV$ for each j, $1 \le j \le 2n+1$.

For R_{n+1} we have

$$R_{n+1}Q^t = \beta \sum_{j=1}^{n-1} \alpha^{2j-1} + 2\alpha^{2n-1} - \beta \sum_{j=0}^{n-1} \alpha^{2j} - 2\alpha^{2n} = DV,$$

as one readily verifies. Observe next that

$$(R_{n+1}-R_n)Q^t = 2q_n - (\nu+1)q_{2n+1}$$
$$= 4\alpha^{2n-1} - (\nu+1)2\alpha^{2n} = 0.$$

For $j = 1$ to $n - 1$,

$$(R_{j+1}-R_j)Q^t = 2q_j - (\nu+1)q_{j+n+1}$$
$$= \beta\alpha^{2j-3} - (\nu+1)\beta\alpha^{2j-2} = 0.$$

Again, for $j = 1$ to $n - 1$ we have

$$(R_{n+j+1}-R_{n+j})Q^t = -(\nu+1)q_j + 2q_{n+j}$$
$$= -(\nu+1)\beta\alpha^{2j-1} + 2\beta\alpha^{2j-2} = 0.$$

Finally,

$$(R_{2n+1}-R_{2n})Q^t = -(\nu+1)q_n + 2q_{2n} + q_{2n+1}$$
$$= -(\nu+1)2\alpha^{2n-1} + 2\beta\alpha^{2n-2} + 2\alpha^{2n},$$

which one readily sees is 0, and we have $R_iQ^t = DV$ for every i, $1 \le i \le 2n+1$. □

Consider next the balanced games where the central diagonal element is -1 and all other diagonal elements are 0. By subtracting adjacent columns we find that necessary and sufficient conditions for a vector P to satisfy

(12.2.1) $\quad PA = K\,(1,1,\ldots,1)$ for some k

are

$$(12.2.2) \qquad \begin{aligned} p_j + p_{j+1} &= (\nu+1)p_{n+j+1} \text{ for } j = 1 \text{ to } n-1; \\ p_k + 2p_{n+1} &= (\nu+1)p_{2n+1}; \\ p_{n+2} &= (\nu+1)p_1; \\ p_{n+j} + p_{n+j+1} &= (\nu+1)p_j \text{ for } j = 2 \text{ to } n. \end{aligned}$$

We rewrite these conditions in the following way:

$$(12.2.3) \qquad \begin{aligned} p_{n+2} &= (\nu+1)p_1, \\ p_2 &= (\nu+1)p_{n+2} - p_1, \\ p_{n+3} &= (\nu+1)p_2 - p_{n+2}, \\ p_3 &= (\nu+1)p_{n+3} - p_2, \\ &\vdots \\ p_n &= (\nu+1)p_{2n} - p_{n-1}, \\ p_{2n+1} &= (\nu+1)p_n - p_{2n}, \\ p_{n+1} &= \tfrac{1}{2}\,[(\nu+1)p_{2n+1} - p_n]. \end{aligned}$$

Proceeding now as in the totally symmetric case [2], we define polynomials

$$(12.2.4) \qquad \begin{cases} F_{-1}(x) = 0, \ F_0(x) = 1, \text{ and} \\ F_k(x) = (x+1)\ F_{k-1}(x) - F_{k-2}(x) \text{ for } k \geq 1. \end{cases}$$

Thus $F_1(x) = x + 1$, $F_2(x) = x^2 + 2x$, etc. By standard difference equations methods we find that the solution of (12.2.4) is

$$(12.2.5) \qquad F_k(x) = (y^{k+1} - y^{-k-1})/(y-y^{-1}),$$

$$\text{where } y = [x+1 + (x^2+2x-3)^{\frac{1}{2}}]/2.$$

Here y and y^{-1} are the two roots of the quadratic

equation $y^2 - (x+1)y + 1 = 0$, and their sum is $y + y^{-1} = x + 1$. It is understood, of course, that if $y = y^{-1}$ then the quotient in (12.2.5) is replaced by a geometric sum.

Since we are interested in making the $F_k(\nu)$ be components of strategy vectors we need to know that they are not negative. For $x \geq 1$ we have $y \geq 1$ and hence $F_k(x) > 0$. For $-3 < x < 1$, y is nonreal and $F_k(x) = 0$ if and only if $y^{2k+1} = 1$ ($y \notin \{1,-1\}$). This holds if and only if $(x+1)/2 = \mathrm{Re}\, y \in \{\cos \frac{h\pi}{k+1}: h = 1,2,\ldots,k\}$. Thus the largest zero of $F_k(x)$ is $x = 2 \cos \frac{\pi}{k+1} - 1$, and we have

(12.2.6) $$F_k(x) > 0 \text{ for } x > 2 \cos \frac{\pi}{k+1} - 1.$$

Now define the 2n+1-component vector P by

(12.2.7) $$P = (F_0, F_2, \ldots, F_{2n-2}, \tfrac{1}{2}F_{2n}, F_1, F_3, \ldots, F_{2n-1}),$$

where $F_j = F_j(\nu)$.

Then each component of P is positive for $\nu > 2 \cos \frac{\pi}{2n+1} - 1$, and in view of (12.2.3) to (12.2.4), P satisfies (12.2.1).

By subtracting adjacent rows instead of columns we find that necessary and sufficient conditions that a vector Q satisfy

(12.2.8) $$AQ^t = K\ (1,1,..,1)^t \text{ for some } K$$

are exactly those expressed in (12.2.2) and (12.2.3) but with the order of the components reversed; i.e., with q_{2n+2-j} in place of p_j. Thus we define Q by

(12.2.9) $$Q = (F_{2n-1}, F_{2n-3}, \ldots, F_1, \tfrac{1}{2}F_{2n}, F_{2n-2}, \ldots, F_2, F_0).$$

It follows that K in (12.2.8) must equal that in (12.2.1) and that the game value is K/D, where D is the sum of the components in P. We summarize these results in the next theorem.

THEOREM 12.3. In the balanced 2n+1 by 2n+1 Silverman game with central diagonal element -1 and all other diagonal elements 0 the optimal strategies for the row and column players, respectively, are P/D and Q/D, where P and Q are given by (12.2.7), (12.2.9) and (12.2.5), and D is the sum of the components of P. The game value is V = K/D, where $K = \sum_{i=0}^{n-1}(F_{2i+1}-F_{2i}) - \frac{1}{2}F_{2n}$.

PROOF. All but the value of K has been proved before stating the theorem. To obtain the value of K we use $K = PC_{n+1}$, where C_{n+1} is the (n+1)-th column of A, and obtain $K = -\sum_{i=1}^{k+1} p_i + \sum_{i=k+2}^{2k+1} p_i$. The asserted value is then immediate. □

For the game with -1 as last diagonal entry and all others 0 we can obtain similar explicit formulas for the column player's optimal strategy vector, but for the row player we have to settle for a rather cyclic kind of recursion which does not seem to yield a similar explicit solution. By subtracting adjacent rows we obtain the conditions

(12.3.1) $$q_i + q_{i+1} = (\nu+1)q_{n+i+1} \text{ for } i = 1 \text{ to } n,$$

$$q_{n+i} + q_{n+i+1} = (\nu+1)q_i \text{ for } i = 1 \text{ to } n-1,$$

and $$q_{2n} = (\nu+1)q_n$$

for the column player's optimal strategy Q. We rewrite these in the form

(12.3.2)
$$\begin{aligned} q_{2n} &= (\nu+1)q_n \\ q_{n-1} &= (\nu+1)q_{2n} - q_n \\ q_{2n-1} &= (\nu+1)q_{n-1} - q_{2n} \\ q_{n-2} &= (\nu+1)q_{2n-1} = q_{n-1} \\ &\vdots \\ q_1 &= (\nu+1)q_{n-2} - q_2 \end{aligned}$$

and $$q_{2n+1} = \frac{1}{(\nu+1)}(q_n + q_{n+1}).$$

Then with the sequence $\{F_k\}$ defined exactly as in (12.2.4) and (12.2.7) we have

(12.3.3) $$Q = \left(F_{2n-2}, F_{2n-4}, \ldots, F_0, F_{2n-1}, F_{2n-3} \ldots, F_1, \frac{1+F_{2n-1}}{\nu+1}\right).$$

By subtracting adjacent columns we obtain the corresponding conditions on the row player's optimal strategy P:

$$\text{(12.3.4)} \qquad p_i + p_{i+1} = (\nu+1)p_{n+i+1} \text{ for } i = 1 \text{ to } n,$$

$$p_{n+i} + p_{n+i+1} = (\nu+1)p_i \text{ for } i = 1 \text{ to } n-1,$$

$$\text{and } \; p_{2n} + 2p_{2n+1} = (\nu+1)p_n.$$

Although these involve the same recursion that we have used to define the polynomials $F_k(x)$ and thereby to obtain explicit formulas for the components of Q here, and of P and Q in the preceding theorems, here there seem to be no clear choices for F_{-1} and F_0 which are independent of n to initialize the process.

THEOREM 12.4. In the balanced 2n+1 by 2n+1 Silverman game with diagonal (0 0 ... 0 -1) the optimal strategy for the column player is Q/D, where Q is given by (12.3.3) and D is the sum of the components of Q. The row player's optimal strategy P is determined by the equations (12.3.4) and $\sum_{i=1}^{2n+1} p_i = 1$.

The game value is

$$V = K/D \text{ , where}$$

$$\text{(12.4.1)} \qquad K = \sum_{j=1}^{n-1} (F_{2j}-F_{2j-1}) + 1 - \frac{1+F_{2n-1}}{\nu+1}.$$

PROOF. All but the value V have been discussed prior to the statement of the theorem. The common value of R_iQ^t, where R_i denotes the i-th row of the payoff matrix, is $R_{n+1}Q^t$, which is seen at once to be K as given by (12.4.1). □

Finally, we can extend the reach of Theorems 12.2 and 12.4 in the following way. (Cf. last paragraph of Section 6.) For any vector W, let W^* denote the vector obtained by reversing the order of the components of W. Let E denote a vector each component of which is 1.

THEOREM 12.5. Let A be the payoff matrix of a balanced Silverman game with diagonal D and game value V. Let A^* be the matrix of the balanced Silverman game with diagonal D^*. If P and Q are vectors with the property that

(12.5.1) $$PA = VE \quad \text{and} \quad AQ^t = VE^t$$

then

(12.5.2) $$Q^*A^* = VE \quad \text{and} \quad A^*P^{*t} = VE^t.$$

Thus in the game A^* the value is V, and Q^* and P^* are optimal strategies for the row and column player, respectively.

PROOF. That (12.5.1) implies (12.5.2) one sees immediately (by writing out the scalar equations if necessary), and the final statement in the theorem follows. □

13. Concluding remarks on irreducibility.

We conclude with brief remarks about the evidence that the reduced games obtained in Sections 8 and 9 are not further reducible. (Those in Sections 10 and 11 clearly are not.)

It is well known that if A is an n by n game matrix with game value V and if $P = (p_1,...,p_n)$ and $Q = (q_1,...,q_n)$ are optimal mixed strategies for the row and column players, respectively, which are completely mixed (i.e., have no zero components), then

$$PA = (V,...,V), \text{ and} \qquad (13.0.1)$$
$$AQ^t = (V,...,V)^t.$$

Moreover, in this case all optimal mixed strategies satisfy (13.0.1). If $V = 0$ and A has rank n-1, or $V \neq 0$ and A has rank n, completely mixed strategies satisfying (13.0.1) are unique optimal strategies, and consequently no optimal strategies exist which are not completely mixed; i.e., the game is not reducible.

Balanced Silverman games with all diagonal elements zero are symmetric, and these are known to be irreducible. The completely mixed optimal strategies are shown in [2] to be unique. We have verified the same in several low order cases for the nonsymmetric reduced games obtained in Sections 8 and 9, when $\nu > 1$. Also, in the course of our studies of these games we have seen machine-generated solutions of hundreds of examples, and without exception the optimal strategies have been completely mixed. We are reasonably confident therefore that these games are not further reducible, but proof of that conjecture must await closer analysis of the rank of these payoff matrices as a function of ν for $\nu > 1$.

(As these notes go to press, the reduced games of Section 8 have been shown to be irreducible when $\nu > 1$, and progress in that direction has been made for those of Section 9.)

References

1. Evans, R.J. Silverman's game on intervals, Amer. Math. Monthly 86 (1979), 277-281.

2. Evans, R.J., and G.A. Heuer. Silverman's game on discrete sets. To appear in Linear Algebra and Applications.

3. Herstein, I., and I. Kaplansky. Matters Mathematical, Harper and Row, New York, 1974.

4. Heuer, G.A. Odds versus evens in Silverman-like games, Internat. J. Game Theory 11 (1982), 183-194.

5. Heuer, G.A. A family of games on $[1,\infty)^2$ with payoff a function of y/x, Naval Research Logistics Quarterly 31 (1984), 229-249.

6. Heuer, G.A. Reduction of Silverman-like games to games on bounded sets. Internat. J. Game Theory 18 (1989), 31-36.

7. Heuer, G.A., and W. Dow Rieder. Silverman games on disjoint discrete sets. SIAM J. on Discrete Mathematics 1 (1988), 485-525.

Vol. 261: Th.R.Gulledge, Jr., N.K. Womer, The Economics of Made-to-Order Production. VI, 134 pages. 1986.

Vol. 262: H.U. Buhl, A Neo-Classical Theory of Distribution and Wealth. V, 146 pages. 1986.

Vol. 263: M. Schäfer, Resource Extraction and Market Structure. XI, 154 pages. 1986.

Vol. 264: Models of Economic Dynamics. Proceedings, 1983. Edited by H.F. Sonnenschein. VII, 212 pages. 1986.

Vol. 265: Dynamic Games and Applications in Economics. Edited by T. Başar. IX, 288 pages. 1986.

Vol. 266: Multi-Stage Production Planning and Inventory Control. Edited by S. Axsäter, Ch. Schneeweiss and E. Silver. V, 264 pages. 1986.

Vol. 267: R. Bemelmans, The Capacity Aspect of Inventories. IX, 165 pages. 1986.

Vol. 268: V. Firchau, Information Evaluation in Capital Markets. VII, 103 pages. 1986.

Vol. 269: A. Borglin, H. Keiding, Optimality in Infinite Horizon Economies. VI, 180 pages. 1986.

Vol. 270: Technological Change, Employment and Spatial Dynamics. Proceedings 1985. Edited by P. Nijkamp. VII, 466 pages. 1986.

Vol. 271: C. Hildreth, The Cowles Commission in Chicago, 1939–1955. V, 176 pages. 1986.

Vol. 272: G. Clemenz, Credit Markets with Asymmetric Information. VIII, 212 pages. 1986.

Vol. 273: Large-Scale Modelling and Interactive Decision Analysis. Proceedings, 1985. Edited by G. Fandel, M. Grauer, A. Kurzhanski and A.P. Wierzbicki. VII, 363 pages. 1986.

Vol. 274: W.K. Klein Haneveld, Duality in Stochastic Linear and Dynamic Programming. VII, 295 pages. 1986.

Vol. 275: Competition, Instability, and Nonlinear Cycles. Proceedings, 1985. Edited by W. Semmler. XII, 340 pages. 1986.

Vol. 276: M.R. Baye, D.A. Black, Consumer Behavior, Cost of Living Measures, and the Income Tax. VII, 119 pages. 1986.

Vol. 277: Studies in Austrian Capital Theory, Investment and Time. Edited by M. Faber. VI, 317 pages. 1986.

Vol. 278: W.E. Diewert, The Measurement of the Economic Benefits of Infrastructure Services. V, 202 pages. 1986.

Vol. 279: H.-J. Büttler, G. Frei and B. Schips, Estimation of Disequilibrium Models. VI, 114 pages. 1986.

Vol. 280: H.T. Lau, Combinatorial Heuristic Algorithms with FORTRAN. VII, 126 pages. 1986.

Vol. 281: Ch.-L. Hwang, M.-J. Lin, Group Decision Making under Multiple Criteria. XI, 400 pages. 1987.

Vol. 282: K. Schittkowski, More Test Examples for Nonlinear Programming Codes. V, 261 pages. 1987.

Vol. 283: G. Gabisch, H.-W. Lorenz, Business Cycle Theory. VII, 229 pages. 1987.

Vol. 284: H. Lütkepohl, Forecasting Aggregated Vector ARMA Processes. X, 323 pages. 1987.

Vol. 285: Toward Interactive and Intelligent Decision Support Systems. Volume 1. Proceedings, 1986. Edited by Y. Sawaragi, K. Inoue and H. Nakayama. XII, 445 pages. 1987.

Vol. 286: Toward Interactive and Intelligent Decision Support Systems. Volume 2. Proceedings, 1986. Edited by Y. Sawaragi, K. Inoue and H. Nakayama. XII, 450 pages. 1987.

Vol. 287: Dynamical Systems. Proceedings, 1985. Edited by A.B. Kurzhanski and K. Sigmund. VI, 215 pages. 1987.

Vol. 288: G.D. Rudebusch, The Estimation of Macroeconomic Disequilibrium Models with Regime Classification Information. VII, 128 pages. 1987.

Vol. 289: B.R. Meijboom, Planning in Decentralized Firms. X, 168 pages. 1987.

Vol. 290: D.A. Carlson, A. Haurie, Infinite Horizon Optimal Control. XI, 254 pages. 1987.

Vol. 291: N. Takahashi, Design of Adaptive Organizations. VI, 140 pages. 1987.

Vol. 292: I. Tchijov, L. Tomaszewicz (Eds.), Input-Output Modeling. Proceedings, 1985. VI, 195 pages. 1987.

Vol. 293: D. Batten, J. Casti, B. Johansson (Eds.), Economic Evolution and Structural Adjustment. Proceedings, 1985. VI, 382 pages. 1987.

Vol. 294: J. Jahn, W. Krabs (Eds.), Recent Advances and Historical Development of Vector Optimization. VII, 405 pages. 1987.

Vol. 295: H. Meister, The Purification Problem for Constrained Games with Incomplete Information. X, 127 pages. 1987.

Vol. 296: A. Börsch-Supan, Econometric Analysis of Discrete Choice. VIII, 211 pages. 1987.

Vol. 297: V. Fedorov, H. Läuter (Eds.), Model-Oriented Data Analysis. Proceedings, 1987. VI, 239 pages. 1988.

Vol. 298: S.H. Chew, Q. Zheng, Integral Global Optimization. VII, 179 pages. 1988.

Vol. 299: K. Marti, Descent Directions and Efficient Solutions in Discretely Distributed Stochastic Programs. XIV, 178 pages. 1988.

Vol. 300: U. Derigs, Programming in Networks and Graphs. XI, 315 pages. 1988.

Vol. 301: J. Kacprzyk, M. Roubens (Eds.), Non-Conventional Preference Relations in Decision Making. VII, 155 pages. 1988.

Vol. 302: H.A. Eiselt, G. Pederzoli (Eds.), Advances in Optimization and Control. Proceedings, 1986. VIII, 372 pages. 1988.

Vol. 303: F.X. Diebold, Empirical Modeling of Exchange Rate Dynamics. VII, 143 pages. 1988.

Vol. 304: A. Kurzhanski, K. Neumann, D. Pallaschke (Eds.), Optimization, Parallel Processing and Applications. Proceedings, 1987. VI, 292 pages. 1988.

Vol. 305: G.-J.C.Th. van Schijndel, Dynamic Firm and Investor Behaviour under Progressive Personal Taxation. X, 215 pages. 1988.

Vol. 306: Ch. Klein, A Static Microeconomic Model of Pure Competition. VIII, 139 pages. 1988.

Vol. 307: T.K. Dijkstra (Ed.), On Model Uncertainty and its Statistical Implications. VII, 138 pages. 1988.

Vol. 308: J.R. Daduna, A. Wren (Eds.), Computer-Aided Transit Scheduling. VIII, 339 pages. 1988.

Vol. 309: G. Ricci, K. Velupillai (Eds.), Growth Cycles and Multisectoral Economics: the Goodwin Tradition. III, 126 pages. 1988.

Vol. 310: J. Kacprzyk, M. Fedrizzi (Eds.), Combining Fuzzy Imprecision with Probabilistic Uncertainty in Decision Making. IX, 399 pages. 1988.

Vol. 311: R. Färe, Fundamentals of Production Theory. IX, 163 pages. 1988.

Vol. 312: J. Krishnakumar, Estimation of Simultaneous Equation Models with Error Components Structure. X, 357 pages. 1988.

Vol. 313: W. Jammernegg, Sequential Binary Investment Decisions. VI, 156 pages. 1988.

Vol. 314: R. Tietz, W. Albers, R. Selten (Eds.), Bounded Rational Behavior in Experimental Games and Markets. VI, 368 pages. 1988.

Vol. 315: I. Orishimo, G.J.D. Hewings, P. Nijkamp (Eds.), Information Technology: Social and Spatial Perspectives. Proceedings, 1986. VI, 268 pages. 1988.

Vol. 316: R.L. Basmann, D.J. Slottje, K. Hayes, J.D. Johnson, D.J. Molina, The Generalized Fechner-Thurstone Direct Utility Function and Some of its Uses. VIII, 159 pages. 1988.

Vol. 317: L. Bianco, A. La Bella (Eds.), Freight Transport Planning and Logistics. Proceedings, 1987. X, 568 pages. 1988.

Vol. 318: T. Doup, Simplicial Algorithms on the Simplotope. VIII, 262 pages. 1988.

Vol. 319: D.T. Luc, Theory of Vector Optimization. VIII, 173 pages. 1989.

Vol. 320: D. van der Wijst, Financial Structure in Small Business. VII, 181 pages. 1989.

Vol. 321: M. Di Matteo, R.M. Goodwin, A. Vercelli (Eds.), Technological and Social Factors in Long Term Fluctuations. Proceedings. IX, 442 pages. 1989.

Vol. 322: T. Kollintzas (Ed.), The Rational Expectations Equilibrium Inventory Model. XI, 269 pages. 1989.

Vol. 323: M.B.M. de Koster, Capacity Oriented Analysis and Design of Production Systems. XII, 245 pages. 1989.

Vol. 324: I.M. Bomze, B.M. Pötscher, Game Theoretical Foundations of Evolutionary Stability. VI, 145 pages. 1989.

Vol. 325: P. Ferri, E. Greenberg, The Labor Market and Business Cycle Theories. X, 183 pages. 1989.

Vol. 326: Ch. Sauer, Alternative Theories of Output, Unemployment, and Inflation in Germany: 1960–1985. XIII, 206 pages. 1989.

Vol. 327: M. Tawada, Production Structure and International Trade. V, 132 pages. 1989.

Vol. 328: W. Güth, B. Kalkofen, Unique Solutions for Strategic Games. VII, 200 pages. 1989.

Vol. 329: G. Tillmann, Equity, Incentives, and Taxation. VI, 132 pages. 1989.

Vol. 330: P.M. Kort, Optimal Dynamic Investment Policies of a Value Maximizing Firm. VII, 185 pages. 1989.

Vol. 331: A. Lewandowski, A.P. Wierzbicki (Eds.), Aspiration Based Decision Support Systems. X, 400 pages. 1989.

Vol. 332: T.R. Gulledge, Jr., L.A. Litteral (Eds.), Cost Analysis Applications of Economics and Operations Research. Proceedings. VII, 422 pages. 1989.

Vol. 333: N. Dellaert, Production to Order. VII, 158 pages. 1989.

Vol. 334: H.-W. Lorenz, Nonlinear Dynamical Economics and Chaotic Motion. XI, 248 pages. 1989.

Vol. 335: A. G. Lockett, G. Islei (Eds.), Improving Decision Making in Organisations. Proceedings. IX, 606 pages. 1989.

Vol. 336: T. Puu, Nonlinear Economic Dynamics. VII, 119 pages. 1989.

Vol. 337: A. Lewandowski, I. Stanchev (Eds.), Methodology and Software for Interactive Decision Support. VIII, 309 pages. 1989.

Vol. 338: J.K. Ho, R.P. Sundarraj, DECOMP: an Implementation of Dantzig-Wolfe Decomposition for Linear Programming. VI, 206 pages. 1989.

Vol. 339: J. Terceiro Lomba, Estimation of Dynamic Econometric Models with Errors in Variables. VIII, 116 pages. 1990.

Vol. 340: T. Vasko, R. Ayres, L. Fontvieille (Eds.), Life Cycles and Long Waves. XIV, 293 pages. 1990.

Vol. 341: G.R. Uhlich, Descriptive Theories of Bargaining. IX, 165 pages. 1990.

Vol. 342: K. Okuguchi, F. Szidarovszky, The Theory of Oligopoly with Multi-Product Firms. V, 167 pages. 1990.

Vol. 343: C. Chiarella, The Elements of a Nonlinear Theory of Economic Dynamics. IX, 149 pages. 1990.

Vol. 344: K. Neumann, Stochastic Project Networks. XI, 237 pages. 1990.

Vol. 345: A. Cambini, E. Castagnoli, L. Martein, P. Mazzoleni, S. Schaible (Eds.), Generalized Convexity and Fractional Programming with Economic Applications. Proceedings, 1988. VII, 361 pages. 1990.

Vol. 346: R. von Randow (Ed.), Integer Programming and Related Areas. A Classified Bibliography 1984–1987. XIII, 514 pages. 1990.

Vol. 347: D. Ríos Insua, Sensitivity Analysis in Multi-objective Decision Making. XI, 193 pages. 1990.

Vol. 348: H. Störmer, Binary Functions and their Applications. VIII, 151 pages. 1990.

Vol. 349: G. A. Pfann, Dynamic Modelling of Stochastic Demand for Manufacturing Employment. VI, 158 pages. 1990.

Vol. 350: W.-B. Zhang, Economic Dynamics. X, 232 pages. 1990.

Vol. 351: A. Lewandowski, V. Volkovich (Eds.), Multiobjective Problems of Mathematical Programming. Proceedings, 1988. VII, 315 pages. 1991.

Vol. 352: O. van Hilten, Optimal Firm Behaviour in the Context of Technological Progress and a Business Cycle. XII, 229 pages. 1991.

Vol. 353: G. Ricci (Ed.), Decision Processes in Economics. Proceedings, 1989. III, 209 pages. 1991.

Vol. 354: M. Ivaldi, A Structural Analysis of Expectation Formation. XII, 230 pages. 1991.

Vol. 355: M. Salomon, Deterministic Lotsizing Models for Production Planning. VII, 158 pages. 1991.

Vol. 356: P. Korhonen, A. Lewandowski, J. Wallenius (Eds.), Multiple Criteria Decision Support. Proceedings, 1989. XII, 393 pages. 1991.

Vol. 358: P. Knottnerus, Linear Models with Correlated Disturbances. VIII, 196 pages. 1991.

Vol. 359: E. de Jong, Exchange Rate Determination and Optimal Economic Policy Under Various Exchange Rate Regimes. VII, 270 pages. 1991.

Vol. 360: P. Stalder, Regime Transitions, Spillovers and Buffer Stocks. VI, 193 pages. 1991.

Vol. 361: C. F. Daganzo, Logistics Systems Analysis. X, 321 pages. 1991.

Vol. 362: F. Gehrels, Essays in Macroeconomics of an Open Economy. VII, 183 pages. 1991.

Vol. 363: C. Puppe, Distorted Probabilities and Choice under Risk. VIII, 100 pages. 1991.

Vol. 364: B. Horvath, Are Policy Variables Exogenous? XII, 162 pages. 1991.

Vol. 365: G. A. Heuer, U. Leopold-Wildburger, Balanced Silverman Games on General Discrete Sets. V, 140 pages. 1991.

Lecture Notes in Economics and Mathematical Systems

For information about Vols. 1–210, please contact your bookseller or Springer-Verlag

Vol. 211: P. van den Heuvel, The Stability of a Macroeconomic System with Quantity Constraints. VII, 169 pages. 1983.

Vol. 212: R. Sato and T. Nôno, Invariance Principles and the Structure of Technology. V, 94 pages. 1983.

Vol. 213: Aspiration Levels in Bargaining and Economic Decision Making. Proceedings, 1982. Edited by R. Tietz. VIII, 406 pages. 1983.

Vol. 214: M. Faber, H. Niemes und G. Stephan, Entropie, Umweltschutz und Rohstoffverbrauch. IX, 181 Seiten. 1983.

Vol. 215: Semi-Infinite Programming and Applications. Proceedings, 1981. Edited by A.V. Fiacco and K.O. Kortanek. XI, 322 pages. 1983.

Vol. 216: H.H. Müller, Fiscal Policies in a General Equilibrium Model with Persistent Unemployment. VI, 92 pages. 1983.

Vol. 217: Ch. Grootaert, The Relation Between Final Demand and Income Distribution. XIV, 105 pages. 1983.

Vol. 218: P. van Loon, A Dynamic Theory of the Firm: Production, Finance and Investment. VII, 191 pages. 1983.

Vol. 219: E. van Damme, Refinements of the Nash Equilibrium Concept. VI, 151 pages. 1983.

Vol. 220: M. Aoki, Notes on Economic Time Series Analysis: System Theoretic Perspectives. IX, 249 pages. 1983.

Vol. 221: S. Nakamura, An Inter-Industry Translog Model of Prices and Technical Change for the West German Economy. XIV, 290 pages. 1984.

Vol. 222: P. Meier, Energy Systems Analysis for Developing Countries. VI, 344 pages. 1984.

Vol. 223: W. Trockel, Market Demand. VIII, 205 pages. 1984.

Vol. 224: M. Kiy, Ein disaggregiertes Prognosesystem für die Bundesrepublik Deutschland. XVIII, 276 Seiten. 1984.

Vol. 225: T.R. von Ungern-Sternberg, Zur Analyse von Märkten mit unvollständiger Nachfragerinformation. IX, 125 Seiten. 1984

Vol. 226: Selected Topics in Operations Research and Mathematical Economics. Proceedings, 1983. Edited by G. Hammer and D. Pallaschke. IX, 478 pages. 1984.

Vol. 227: Risk and Capital. Proceedings, 1983. Edited by G. Bamberg and K. Spremann. VII, 306 pages. 1984.

Vol. 228: Nonlinear Models of Fluctuating Growth. Proceedings, 1983. Edited by R.M. Goodwin, M. Krüger and A. Vercelli. XVII, 277 pages. 1984.

Vol. 229: Interactive Decision Analysis. Proceedings, 1983. Edited by M. Grauer and A.P. Wierzbicki. VIII, 269 pages. 1984.

Vol. 230: Macro-Economic Planning with Conflicting Goals. Proceedings, 1982. Edited by M. Despontin, P. Nijkamp and J. Spronk. VI, 297 pages. 1984.

Vol. 231: G.F. Newell, The M/M/∞ Service System with Ranked Servers in Heavy Traffic. XI, 126 pages. 1984.

Vol. 232: L. Bauwens, Bayesian Full Information Analysis of Simultaneous Equation Models Using Integration by Monte Carlo. VI, 114 pages. 1984.

Vol. 233: G. Wagenhals, The World Copper Market. XI, 190 pages. 1984.

Vol. 234: B.C. Eaves, A Course in Triangulations for Solving Equations with Deformations. III, 302 pages. 1984.

Vol. 235: Stochastic Models in Reliability Theory. Proceedings, 1984. Edited by S. Osaki and Y. Hatoyama. VII, 212 pages. 1984.

Vol. 236: G. Gandolfo, P.C. Padoan, A Disequilibrium Model of Real and Financial Accumulation in an Open Economy. VI, 172 pages. 1984.

Vol. 237: Misspecification Analysis. Proceedings, 1983. Edited by T.K. Dijkstra. V, 129 pages. 1984.

Vol. 238: W. Domschke, A. Drexl, Location and Layout Planning. IV, 134 pages. 1985.

Vol. 239: Microeconomic Models of Housing Markets. Edited by K. Stahl. VII, 197 pages. 1985.

Vol. 240: Contributions to Operations Research. Proceedings, 1984. Edited by K. Neumann and D. Pallaschke. V, 190 pages. 1985.

Vol. 241: U. Wittmann, Das Konzept rationaler Preiserwartungen. XI, 310 Seiten. 1985.

Vol. 242: Decision Making with Multiple Objectives. Proceedings, 1984. Edited by Y.Y. Haimes and V. Chankong. XI, 571 pages. 1985.

Vol. 243: Integer Programming and Related Areas. A Classified Bibliography 1981–1984. Edited by R. von Randow. XX, 386 pages. 1985.

Vol. 244: Advances in Equilibrium Theory. Proceedings, 1984. Edited by C.D. Aliprantis, O. Burkinshaw and N.J. Rothman. II, 235 pages. 1985.

Vol. 245: J.E.M. Wilhelm, Arbitrage Theory. VII, 114 pages. 1985.

Vol. 246: P.W. Otter, Dynamic Feature Space Modelling, Filtering and Self-Tuning Control of Stochastic Systems. XIV, 177 pages. 1985.

Vol. 247: Optimization and Discrete Choice in Urban Systems. Proceedings, 1983. Edited by B.G. Hutchinson, P. Nijkamp and M. Batty. VI, 371 pages. 1985.

Vol. 248: Plural Rationality and Interactive Decision Processes. Proceedings, 1984. Edited by M. Grauer, M. Thompson and A.P. Wierzbicki. VI, 354 pages. 1985.

Vol. 249: Spatial Price Equilibrium: Advances in Theory, Computation and Application. Proceedings, 1984. Edited by P. T. Harker. VII, 277 pages. 1985.

Vol. 250: M. Roubens, Ph. Vincke, Preference Modelling. VIII, 94 pages. 1985.

Vol. 251: Input-Output Modeling. Proceedings, 1984. Edited by A. Smyshlyaev. VI, 261 pages. 1985.

Vol. 252: A. Birolini, On the Use of Stochastic Processes in Modeling Reliability Problems. VI, 105 pages. 1985.

Vol. 253: C. Withagen, Economic Theory and International Trade in Natural Exhaustible Resources. VI, 172 pages. 1985.

Vol. 254: S. Müller, Arbitrage Pricing of Contingent Claims. VIII, 151 pages. 1985.

Vol. 255: Nondifferentiable Optimization: Motivations and Applications. Proceedings, 1984. Edited by V.F. Demyanov and D. Pallaschke. VI, 350 pages. 1985.

Vol. 256: Convexity and Duality in Optimization. Proceedings, 1984. Edited by J. Ponstein. V, 142 pages. 1985.

Vol. 257: Dynamics of Macrosystems. Proceedings, 1984. Edited by J.-P. Aubin, D. Saari and K. Sigmund. VI, 280 pages. 1985.

Vol. 258: H. Funke, Eine allgemeine Theorie der Polypol- und Oligopolpreisbildung. III, 237 pages. 1985.

Vol. 259: Infinite Programming. Proceedings, 1984. Edited by E.J. Anderson and A.B. Philpott. XIV, 244 pages. 1985.

Vol. 260: H.-J. Kruse, Degeneracy Graphs and the Neighbourhood Problem. VIII, 128 pages. 1986.

continuation on page 141